AF821970

SOTHIS PRESS

Sothis Press aims to completely reimagine the public domain book with an emphasis on universal access to captivatingly designed print editions. We believe that technology will enable every single book ever published to be perpetually accessible in beautiful print form. Sothis is beauty in ubiquity.

www.sothis.press
1 Waterhouse Square, London, ECIN 2ST

Nine Motorboats and How to Build Them

A Book of Complete Building Plans and Instruction, Which Contains All Necessary Information for the Amateur Who Wants to Build His Own Boat

Motor Boat Publishing Company

SOTHIS PRESS

ISBN 978-1-0365-0517-2

Nine Motorboats and How to Build Them
is in the public domain.

This edition is published by
Sothis Press © 2023. The company
is registered in England and Wales.
The Company Number is 14900389.

NINE MOTORBOATS

AND

HOW TO BUILD THEM

Second Edition

A Book of Complete Building Plans and Instruction, which contains all necessary information for the amateur who wants to build his own boat.

NEW YORK
THE MOTOR BOAT PUBLISHING COMPANY
110-112 WEST 40th STREET

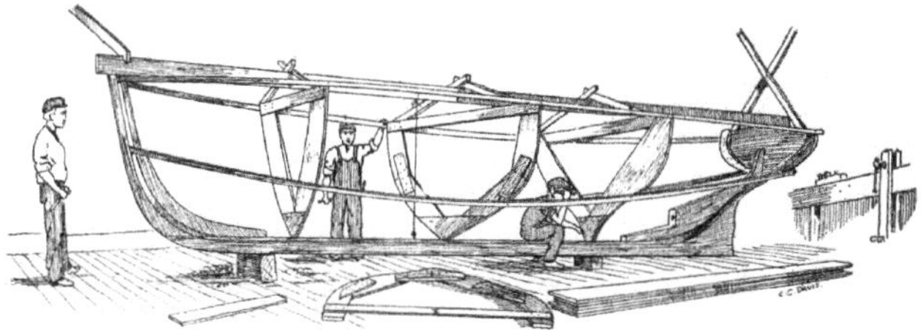

CONTENTS

"Pop-Gun"—A Twenty-Foot Runabout

DESIGNED BY C. G. DAVIS

THIS little runabout, named *Pop-Gun* by one who saw a model of her, is a 20-footer of such simple construction that anyone can undertake her construction with certainty of success—that is, anyone who knows enough of the use of tools to build a box. All that is needed is a knowledge of how to use a saw, how to plane the edge of a board straight, and how to nail lumber together. There are very few curved cuts to be made, and nearly every cut is along a straight line that may be snapped with a chalk-line or drawn with a straight edge. In fact, the designing of a boat like this is the hardest part, and that is done.

Simplicity in every feature has, of course, been the leading object in designing *Pop-Gun*, but simplicity is quite possible in a very good little runabout, and *Pop-Gun* will prove a handy, serviceable, sturdy boat that will carry several people easily, and travel along at about eight miles an hour with a 3-hp. motor.

The construction of a round-bottomed boat requires considerable skill other than that necessary to use the tools. The shaping of each frame and plank is quite an art; but in this boat the construction is so simple that anyone should succeed in producing a good boat. We will take each piece of wood in turn and tell you just how to shape it.

First come the three patterns, or molds, that give you the desired shape. These are shown clearly and with all the necessary dimensions marked on them for reproducing them full size. A large sheet of brown wrapping paper will do if a clean board floor is not available to draw them on.

Take the shape of the middle mold, No. 2, for example. Draw a straight line horizontally across the paper or floor and with a large, steel carpenter's square draw a center line, which we show here dotted at right angles to it. Each side of this center line measure off 2 feet 4½ inches, which represents the width of the boat at the deck to the inside of the planking. Eighteen and one-quarter inches below this—the distance given in the plans of the molds—draw another horizontal line and measure out two feet each side—that distance, 18¼ inches, represents the vertical depth of the side plank, which of course, if measured on the angle will be a fraction longer; 19 inches it really is. Four and three-quarter inches below the second line draw another short one representing the keel and measure out three inches each side, the keel being six inches wide. By drawing lines connecting these spots you have the outline of the mold. The others are found in exactly the same way, using the distances marked on the plans. Wooden patterns have to be made of these three molds so that when properly spaced and set up the planks forming the boat can be bent around them. This causes quite a little strain to come on the molds, so do not build them so flimsily that they will give or break under the strain, and cause your boat to be built crooked.

In Figure 1 I have shown one method of putting together a mold, and in Figures 2 and 3 other ways are shown. It matters little which way you build them; the

FIG-4.

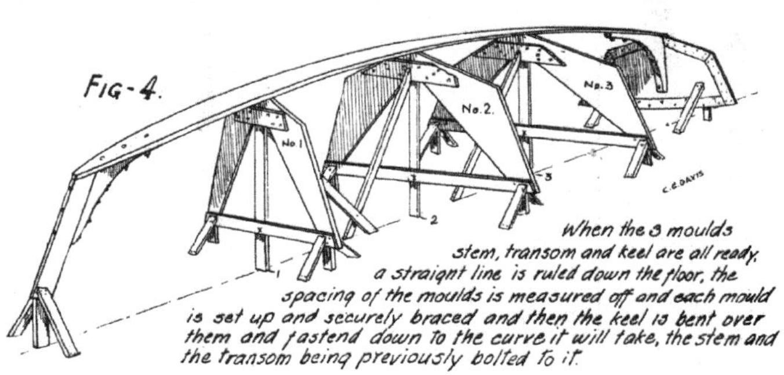

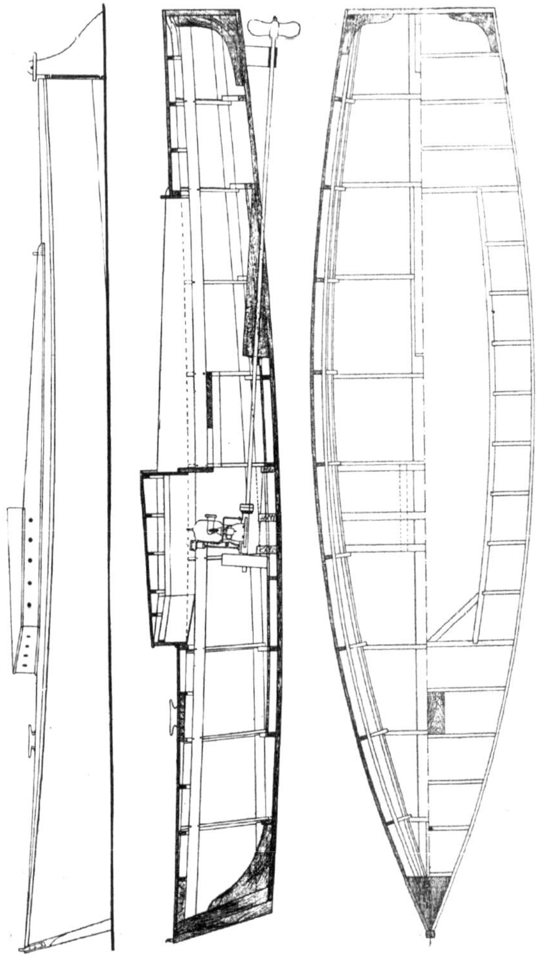

CONSTRUCTION PLANS OF "POP GUN"

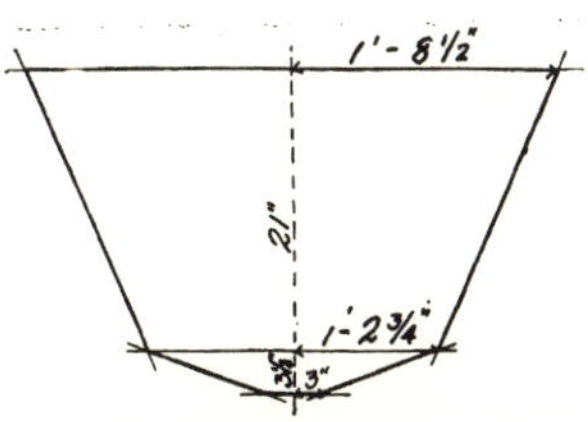

Mould No-1.

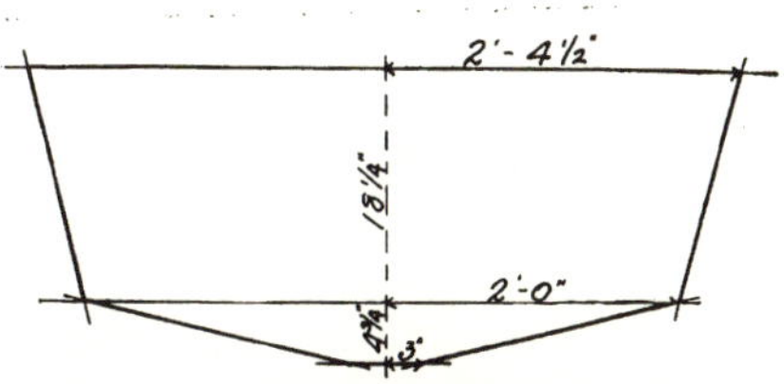

Mould No-2.

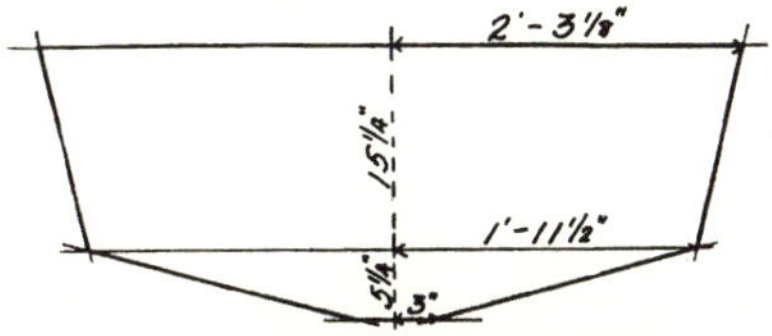

Mould No-3

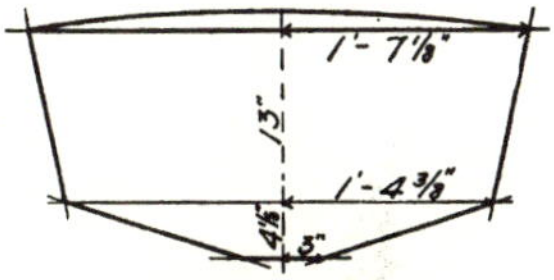

Transom.

one that cuts up your available wood to best advantage is the one to use. Cleat the various pieces securely together and nail a brace about 4 inches wide by 1 inch thick across the heads of each mold. Mark the center line at top and bottom to assist you when you come to set the molds up along a center line. Don't cut up expensive wood for molds, for they are only temporary affairs that will be thrown away when the boat is built.

The transom is to be cut out of 1¼-inch oak to the size and shape given. Try to get good dry, seasoned oak and, if possible, a piece wide enough to make the transom in one piece. If you can't get this, make it of two pieces but keep the seam between the two well up, so that it will be out of water most of the time, and therefore not be liable to leak. Saw it out to the given shape with square edges. The bevels necessary to let the plank lay flat on it can be planed off later, when it is set up, by bending a board over the molds so that it touches the transom and then planing off the after edge until the plank lays flat on the transom. (Figure 4.)

If the transom must be made of two pieces, plane up the two edges that are to meet so that they make a perfect fit. You can soon tell whether they do or not by holding them up to the light of a window or lamp (Figure 5) and the light shining through the crack of the seam will soon show you where the high spots are. Mark them and run a shaving off, and then try again. Keep at it till you make a perfect fit. The amateur is very apt to try to do this with a short plane; most amateurs show a preference for the smoothing plane but if they will take a long "jointer" plane they will get better results.

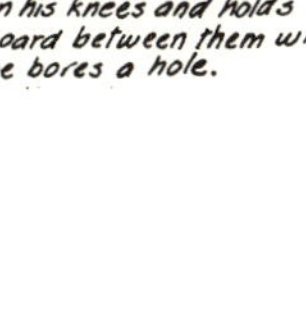

The amateur generally has no bench vise, he gets down on his knees and holds the board between them while he bores a hole.

A strong, neat joint can be made by dowelling these two together. Lay the two pieces of the transom flat on floor and rule about four straight lines across both; but be sure to have them square to the seam and parallel with each other (Figure 6), not staggered, as in Figure 7. If they are put, as shown in Figure 7, you can never drive the two together as the dowels will bind and prevent this. The holes must be very carefully bored so that they will not run and come out through the side of the board. Clamp one piece of the transom in a bench vise if you are using one or get down on your knees, as nine-tenths of the amateurs have to, and hold it between your knees. By looking squarely down on it you can tell pretty

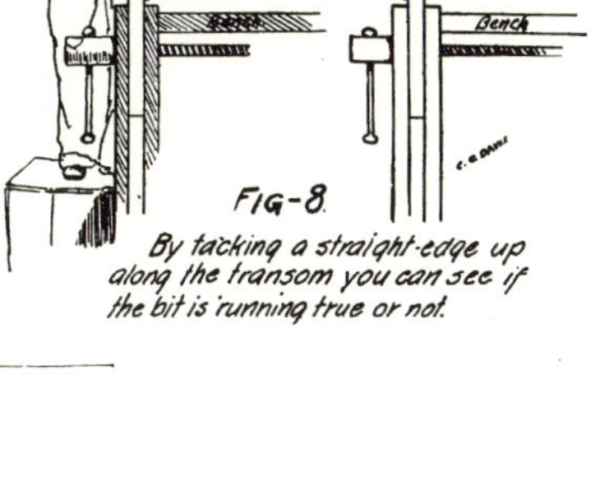

FIG-8

By tacking a straight-edge up along the transom you can see if the bit is running true or not.

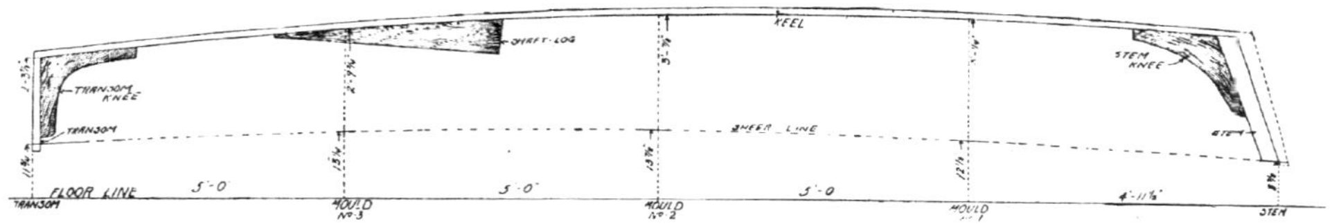

MEASUREMENTS NECESSARY FOR SETTING UP THE KEEL

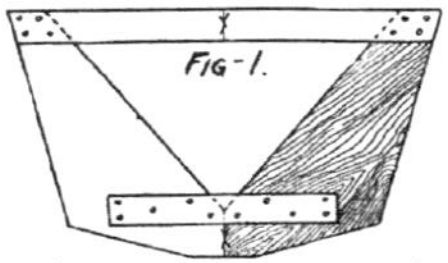

If you have some wide boards the mould can be made of two pieces strapped together at the bottom

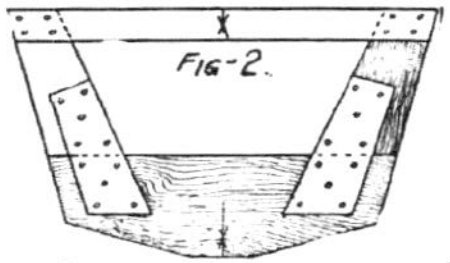

One wide board can go across the bottom and two side pieces cleated to it

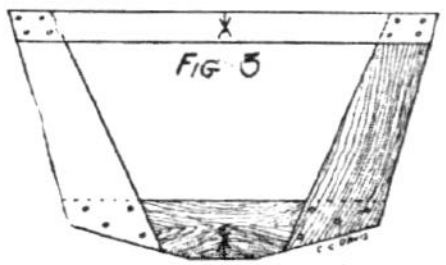

Instead of cleating the side pieces they can lap and be screwed or nailed fast

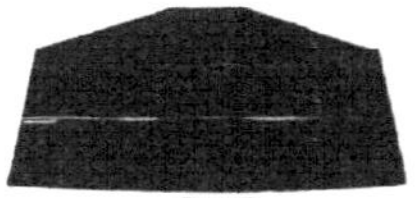

FIG-5

By holding the two pieces of the transom up to a strong light you can see where the high spots are and know where to plane off

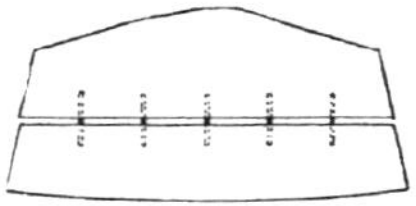

FIG-6

In doweling the two pieces of a transom together be sure all the dowels are parallel to each other

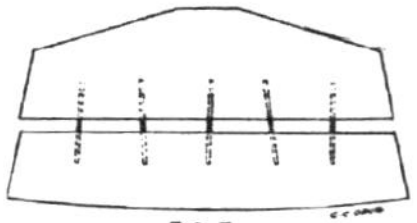

FIG-7

Do not be careless and let the dowel stand all different angles or you will never be able to drive them together, they will bind

accurately whether the bit is running true or not. A good way to get the bit started true is to tack a straight-edged stick against the face of the board opposite the hole. With this guide you can keep the shank of the bit true by your eye. Figure 8 illustrates this point clearly.

Bore in to a depth of about four inches into each side. Don't try to make the bit cut too fast—just press on the handle of the bit with one hand and after you have cut in a couple of inches pull the bit out and run your hand down it to sweep the shavings out of the worm of the

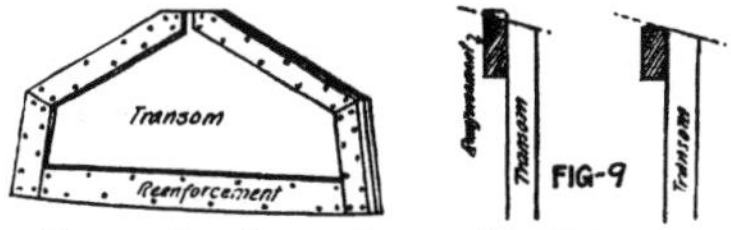

Showing how the reenforcement is fastened around the edge of the transom to give a broader nailing surface for the sides and bottom boards.

screw. If you feel the steel is quite hot you know you are crowding the bit too fast. Crowding too hard is what causes the bit to run out to one side and may cause the point of it to come out through the side, disfiguring your transom. Some people lay on a brace and bit and put all their weight into it and then wonder why the bit gets bent or cuts crooked. Give the cutting point time to do its work. A bit isn't a conductor's punch ,to be pushed through by force—turn it. Dip the ends of the dowels into some liquid glue before you drive them into the transom.

In order to give a better nailing surface to the side and bottom planks, take some pieces of the 1¼-inch oak about three inches wide and screw it fast all around the edge of the transom. Use about 1¾-inch brass screws and bore so that their heads go in about ⅜ of an inch. Set

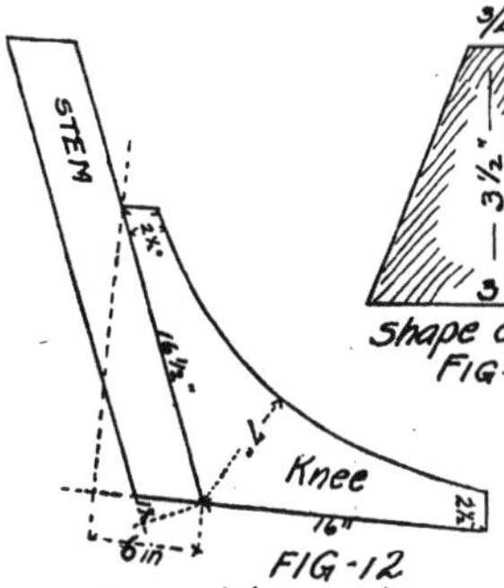

FIG-12
Shape of knee at bow.

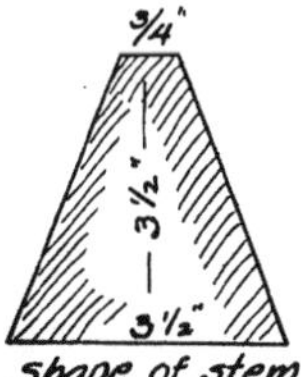

Shape of stem.
FIG-10

these re-enforcing pieces so they extend out beyond the edge of the transom so that when you come to bevel the transom to take the sides and bottom planks there will be wood enough to do so and it will not be as in Figure 9.

Round up the top edge of the transom two inches higher in the center than it is at the sides to give the necessary crown to the deck. The keel is a parallel oak board, twenty feet long, six inches wide and 1½ inches thick that one can have sawed for him at the lumber yard where he buys the wood. Be sure to get a sound, clear oak board.

The stem is built just as ninety-nine out of a hundred skiffs are built—of two pieces, and this greatly simplifies the work. There is no rabbet to be cut. Just get a piece of dry, sound oak, 28 inches or more in length and 3½ inches square. Twenty-eight inches is the exact, neat length required, but if you get it out 3 feet long it will be long enough to reach the floor so you can brace it securely and then saw it off after you turn the boat over. Down the side that is to be the front edge draw a center line and ⅜ of an inch each side of it draw lines parallel to it; this represents the ¾-inch face that is to be left square, as shown in figure 10. With draw-knife and plane cut this stem to the wedge shape shown.

To secure the transom and stem to the keel, cut out two knees of 3-inch oak or hackmatack so that the grain follows, in a measure, the shape of the knees and is not so crossgrained as to split off easily. Cut them accurately to the angles given in Figures 11 and 12, and be sure to have the edges square and true so that when the stem is riveted to the keel they will stand square and not be off to one side, or crooked. If you have ever done any riveting these knees can be riveted to the keel, stem and transom by some ⅜-inch galvanized iron rod, riveted over clinch rings. If you have not done any of this kind of work, I would advise bolting them fast with galvanized iron carriage bolts, putting washers under the nuts. Set the heads in just flush with the outside of the keel and stem, but at the transom let the heads in far enough (⅜ of an inch will do) to allow a wooden plug being inserted to hide the bolt head. Dip the bolt into varnish or paint it with red lead if you cannot get the galvanized iron bolts and are forced to use black iron.

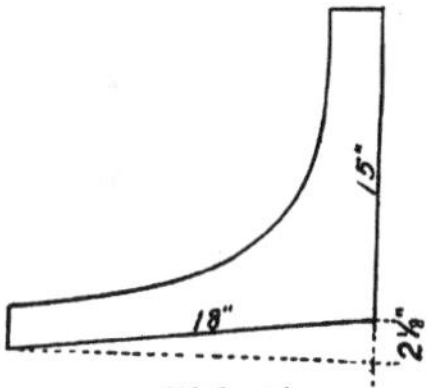

FIG-11
Shape of knee that holds transom to keel.

You will find that now is the time to bore the shaft hole through the keel, as you can see both inside and outside of the boat, and line up the angle you want your shaft to run with a chalk line stretched along the edge of the keel plank. Get out a shaft log of clear, sound, 4-inch oak and fit it to the under side of the keel as it is bent over the molds. To do so you will have to cut out part of the mold but that is of no consequence.

When you have scribed and cut this shaft log so that it makes a perfect fit, lay a piece of heavy muslin painted with rather thick white lead paint between the log and keel and with boat-builders' screw-clamps screw the log firmly in place until you have bored and bolted it fast with a row of carriage bolts along each side. Take care to leave the center clear of bolts so that the shaft hole can be bored through.

To bore this hole looks like a momentous question to the amateur, and if he tries to do it without proper provision beforehand he will find that it is difficult. First of all, get an auger of the proper diameter for the hole you want to bore. This diameter varies with the engines and depends upon what make of engine you decide to instal.

Take that auger to a blacksmith and have him weld on a long handle, five feet from the end of the auger to the

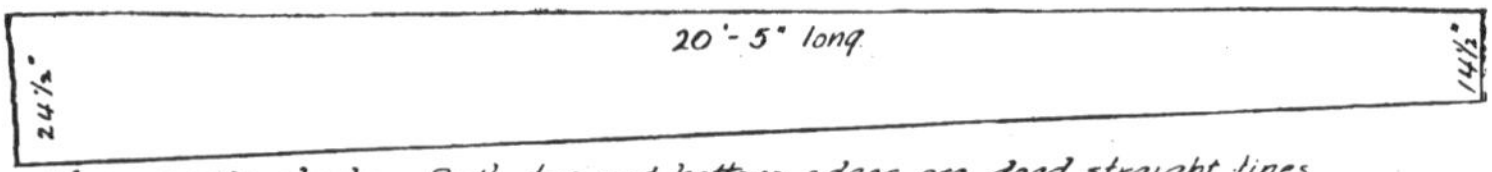

Plan of side planks Both top and bottom edges are dead straight lines.

crank, and let him bend a crank to turn it, with 7 inches throw.

You cannot bore from the inside out and to start the bit on the outside, clamp on, temporarily, a block of oak just so that the bit can start into it and get cutting true before it strikes the slanting keel. By so starting the bit will bore clean and true through the keel and so on through the deadwood, or shaft log.

To hold the auger true while starting to bore, rig up a couple of temporary uprights with a score cut in them just enough to steady the auger or bit as shown in Figure 13.

With this much done, the troublesome features of the construction are practically over and the rest is plain sailing. The sides, in localities where wide lumber is obtainable, can be made of one board to a side. A ¾-inch cedar or pine board 20 feet 6 inches long, 24½ inches wide forward and 14½ inches wide aft will do. Both top and bottom edges of this side plank are dead straight lines, requiring no shaping whatever, which makes the work of building such a boat much easier for a novice. He can lay out his plank with a chalk line or straight edge, and so long as the planks are 24½ inches wide at one end and 14½ inches at the other in a distance of 20 feet 6 inches, he cannot go wrong. By bending such a plank around the molds you can soon see how the edge of the transom will have to be beveled so that it fits flat. When that little cutting has been done, proceed to fasten the side planks into place. This part of the work is very interesting to the amateur, as in a short time he sees the boat rapidly take shape and begin to look like a real boat.

If the side planks cannot be gotten in one width they can be made of two narrower ones. The frames that are fitted in afterwards will hold them securely together and the seam can be caulked and made water-tight. This seam will be above the water level, however, so it is not likely to leak.

Let the side planks extend an inch or so beyond the stem and transom and saw them off flush afterwards. Fasten the side planks with either 1¼-inch galvanized boat nails or 1½-inch brass screws to the stem and transom. To make a neat job the heads of the nails or screws should be counter-sunk about ¼ or 3-16 of an inch and the hole either filled with putty or a wooden plug.

The lower edge of the side plank must have an oak batten along its edge to give a greater nailing surface for the bottom boards. To put this batten in, saw a notch out of the corner of the molds and then bend in the batten, which should be of oak about ⅞-inch thick and 2¼ inches wide. Either rivet this to the edge of the side plank or screw it fast from the inside, letting the edge extend down far enough so that it can be beveled off to take the bottom boards, just as the re-enforcement for the transom was put on.

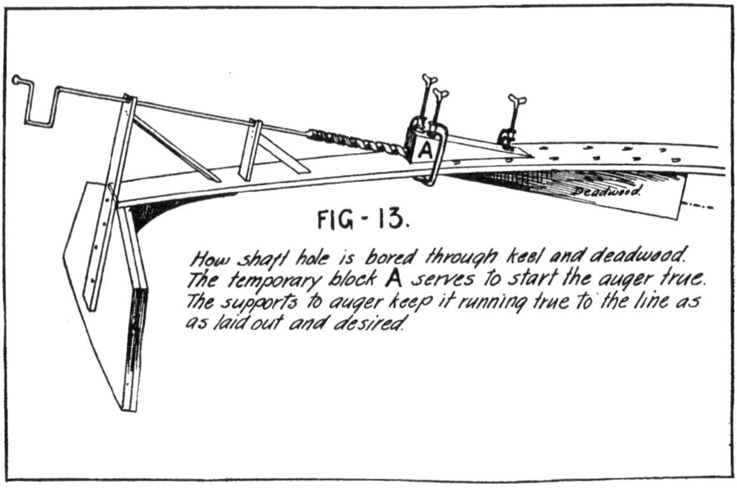

FIG-13.

How shaft hole is bored through keel and deadwood. The temporary block A serves to start the auger true. The supports to auger keep it running true to the line as as laid out and desired.

A "Pop Gun" built by Smith Kennerson of Rochester, N. Y.

PART II

THE next step in the construction of *Pop-Gun* is to get out a lot of strips of oak for frames ⅞-inch by 2 inches wide and rivet them across the inside of the side planks in the form of frames at intervals of every two feet. You can rivet them in or nail them fast from the outside, setting the nail heads in for either putty or wooden plugs.

At each up-and-down-side frame so riveted in on the side planks, fit a floor frame of the same size so that it notches into the chine piece alongside the frame and is nailed diagonally through it into the chine piece, the other end beveled so that it fits flatly on the inside of the keel, the two butted together in the center. Screw the end of the frame fast to the keel and when all the floor frames

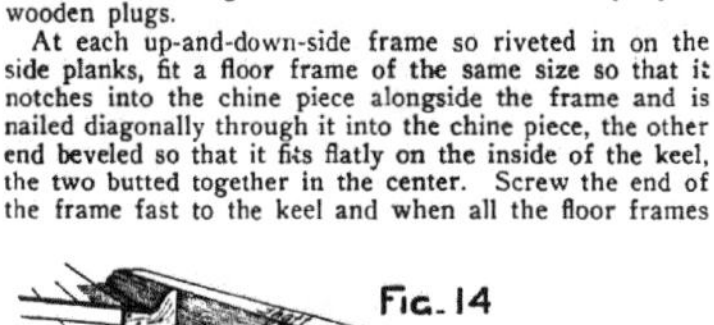

are in fit in some 1-inch thick sawed floors as shown in the midship section. Fit them in alongside of the frames and nail them fast to the frames and keel.

Where the deadwood, or shaft log, is riveted fast to the keel the floor frames cannot, of course, go across the boat, and to secure them at this point various means are resorted to. Some builders cut a dovetail jog in the shaft log and dovetail heels of the frames into it; others simply butt the frame up against the log and rivet it to the keel; but a better way than either of these is to cut out some small oak knees and rivet them alongside of the frames and screw them fast to the shaft log, as in Figure 14.

To carry the engine a couple of heavy floors are fitted across high enough to fit onto the chine pieces and securely riveted or bolted to them. By being fastened to the chine pieces and keel in this way they distribute the strain of the engine over sufficient area to prevent its straining the hull in any way. Be sure to cut a groove or limber hole, as it is called, across the underside of all floors down near the keel to allow any bilge water that may collecet to run aft, where it can be pumped out.

When all these floors are securely fastened in you are ready to plank the bottom of the boat. For this you need some ⅞-inch cedar boards about 9 inches wide and 19 feet long.

The first plank each side of the keel, called the garboard, will have to have a slight bevel planed on its edge next to the keel, so that the seam will not stand open too wide. It must be wider open on the outside than the inside, so that the caulking will wedge in tight as it is driven in, but not so wide open as a square edge on the board would make it.

You will find a straight plank will, when it is put on over the floor frames, touch the keel forward and aft, and amidships it will stand away 1¾ inches, but the plank can be shoved in edgeways, so that it will fit tightly all along the keel.

The second plank on the bottom will not be quite so long and can be edge-sprung the same as the garboard was, and a third plank on each side will complete the bottom. Cut the edges of each board so it is flush with the side plank and either screw fasten or nail them to the edge of the side planks and frames. If you have bought clear cedar plank you will not have much plugging up of knot holes to do, but absolutely clear cedar is almost impossible to get. Try all knots that look loose and knock them out, filling the hole by driving in a pine or a cedar plug and sawing it off flush outside.

The loose knots are usually distinguishable by a fine black ring of bark around them. Ream out all of the black stuff so that clear wood shows, for if you leave it, it will only rot away and your plug will then be loose and liable to come out.

Plane off the seams so that they are smooth and even, and with medium sandpaper folded over a block of wood just large enough to be held in the hand easily, scrub crossways across the grain and then with fine sandpaper rub it fore and aft with the grain. Before you give it

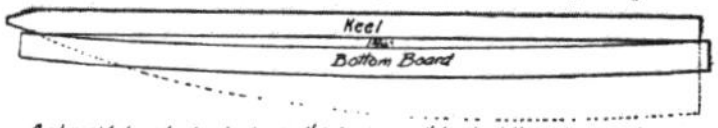

A straight board when bent over the frames will touch at the ends and show an opening of 1¾ inches in the middle, but the board can be pushed up tight.

the final rubbing, the bottom planks should be caulked. The amount of cotton necessary for this will vary with the size of the seam you have left between the boards. The seam should be perfectly tight inside and opened about an eighth of an inch on the outside, but some of our amateur builders may be dismayed to find a plank that was put on perfectly tight has, at the end of a few days, shrunk so that it is opened a sixteenth of an inch or so. It is for this reason that the seam is made wider open on the outside, so that the cotton you are caulking with jams tighter and tighter as it is hammered into the wedge-shaped seam. But if the seam is wrongly beveled, that is, so that the seam is wider open on the inside than the outside, you can readily see you cannot keep the cotton from falling through, and as all the pressure is on the

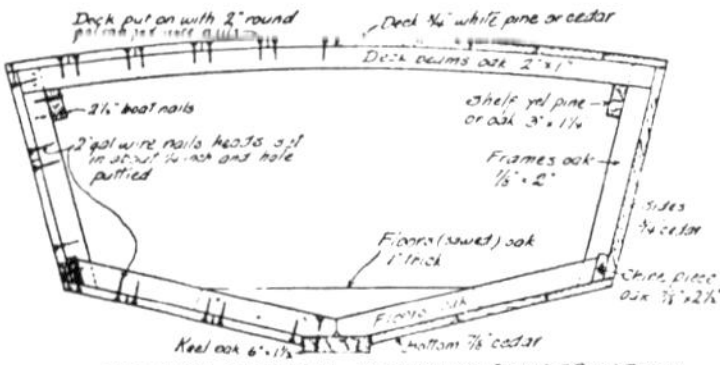

MID-SHIP SECTION SHOWING CONSTRUCTION.

outside, if it is beveled as first described you can see it is being pushed in against the bevel of the plank.

The seam is not apt to be open the same amount all along and to even this up somewhat various methods are resorted to. One way is to take a sharp-edged caulking iron (Figure 15) and drive it in the same distance all along to even the seam. Another way is to make a "beetle" out of hard white oak. A beetle is nothing more than a round or oval piece of oak about 4 inches long, 2 inches diameter if round or about 1½ by 2 inches if oval, with a wedge-shaped tongue cut across the long way of it, as shown in Figure 16. By being cut on a bevel as shown, you can insert this beetle in one end of a seam, and by driving on the beetle with a mallet send it along the whole length of the seam, making it of uniform width and bevel, as the cedar will readily compress when the beetle comes to a narrow place. Another way, but one that requires metal wheels, is to have beveled edge wheels set in a handle about 15 inches long (Figure 17). By rolling this wheel along the seam it can be opened in a few moments.

The cotton used in caulking boats comes rolled up in a ball already spun into a band about as thick as your thumb, but that, you will find, will readily split to any desired thickness. Take off a string of it large enough to pack tightly into the seam and either hammer it in with a *blunt* caulking iron or roll it in with a *blunt*-edge roller; the sharp ones you used to open the seam will cut the cotton all up. Where spun boat cotton cannot be obtained take the ordinary cotton and pull it out into a sort of string and roll it under the palm of your hand over your knee. You will find it is very apt to break apart, but by simply laying the ends together and rolling

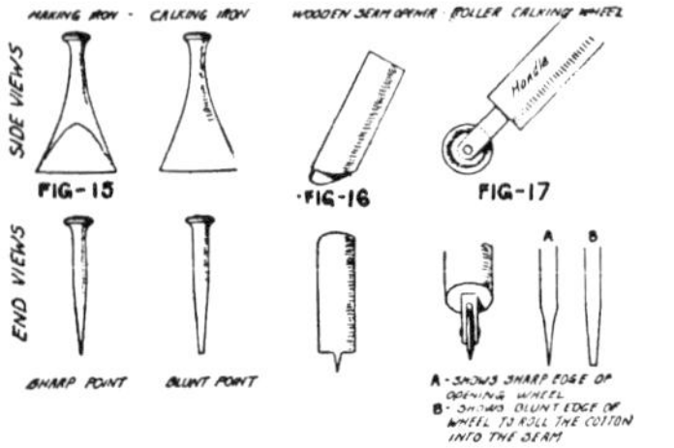

them they will unite again. On very small seams cotton wicking that can be purchased in balls is used. This consists of several cotton cords loosely twisted into one round wick, but any number of these cords can be separated. Take out three, four, five, or as many strings as will be necessary to fill the seam and twist them up a little before you hammer or roll them into the seam.

In caulking, where you come across a hole larger than the general run of the seam bunch the cotton up as shown in Figure 18 by looping it into loops. When these are hammered into the seam it will fill the hole.

The caulking should all be driven in so that it is at least an eighth of an inch or so below the edges of the planks to give room for the putty. To hold the cotton into the seam and to form a holding-on for putty, paint the seam over the cotton. You can do this with a narrow chisel-pointed brush, but the edges of the boards will get as much paint as the seams. Take a handful of shavings or a piece of cloth and wipe off the surplus paint if you have to use such a brush. There is a special brush made for this purpose, called a seam brush, consisting of a single row of bristles set in a thin wooden handle. When this paint is dry, mix up some putty and putty all the seams and nail heads or screw heads not covered by wooden plugs and then paint the bottom with some anti-fouling paint. Green looks very pretty with a white top side.

When the bottom is dry, knock all the braces loose from the floor and molds, transom and stem. Get a couple of men to help you and pick the boat up by the ends and turn her over right side up onto a couple of boxes or heavy timbers.

The bottom being complete the next step is to build the frame for the deck. For a shelf to lay these deck beams

FIG - 18.

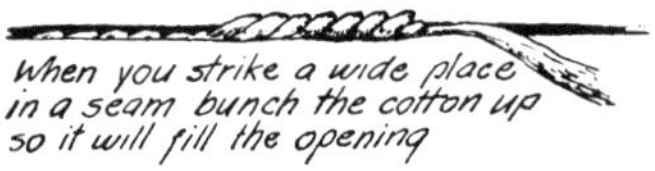

on you want two strips of oak—yellow pine will do if oak is hard to get—1¼ inches thick, 3 inches wide and 20 feet long. Nail these fast to each of the frames so that the upper edge is about 1½ inches below the level of the edge of the side planks. If there were no round or crown, as it is called, to the deck beams, which are 2 inches deep and 1 inch thick, this shelf would be put just 2 inches below the side line, but to allow for the arch-up of the deckbeam, which is about ½ inch at the inner edge of the frames, the shelf is raised that much so the underside of the deck beam will rest on it and the top of the deck beam will come even with the top of the side plank and not half an inch or so below it, as in Figure 19.

To get out the oak deck beams you must first make a pattern of the curve wanted. A flat deck made of straight deck beams spoils the looks of any boat. In your haste to get the boat done do not be tempted to use flat deck beams.

I have shown a crown of 2½ inches in 5 feet on the plans. You can sweep such a curve by taking a batten 15 feet long, drive a brad through one end as a pivot and hold a pencil at other end as you sweep an arc of a circle on the floor or on a thin board you are going to cut as a pattern. The measurements of such a curve are given in Figure 20.

With this pattern you can mark out as many beams as you want and saw them out by hand with a rip-saw, planing them up smooth afterwards. If you have no circle plane to go around on the inside of the curve, clean it with a spoke shave. Get out two beams of 1½ inches in thickness, one to go across at the forward end of the cockpit and one at the after end. The first is fitted in alongside of the third frame from the stem, the other at the second frame from the stern.

In the general construction plan one style of deck framing is shown that requires considerable cutting and fitting, but a far more simple way is illustrated in Figure 21.

The sill piece is of 1½-inch wide by 2 inches deep oak or yellow pine—oak if the beams are to be jogged into it, yellow pine will do if not. In Figure 21 there is no cutting to do; the sill piece is bent parallel with the side of the boat one foot in from the outer edge and riveted to the underside of two beams forward and two aft. Brace it up temporarily until all the short beams are cut and riveted into their places, as shown on the plans. Then fit one permanent brace, about midway of its length, and fasten it to a crossfloor below to stiffen the deck, should anyone sit on it along the sides.

Nail the beams down into the shelf with 3-inch galvanized wire nails. Fit a V-shaped block of oak across from side to side away up in the extreme bow, resting on top of the shelves and butting against the after side of the stem, thick enough (about 2½ inches) to allow you to cut the proper crown of the decks on it so the decks will lay flat on it. This not only stiffens the boat but gives you wood to screw a flagpole socket or bevel low chocks fast to.

Fit another similar block between the beams of the forward deck as shown on the plans to receive the screws holding a cleat, and aft in the corner formed by the side and the transom fit a small knee to stiffen it.

The deck should be either of ¾ or ⅝-inch pine or cedar put on in wide boards and painted or varnished as suits one's taste. I would not advise having the boards too wide;

FIG-20.

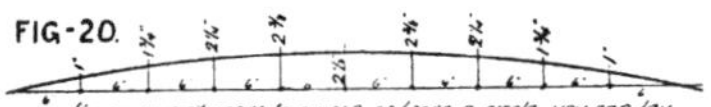

If you have nt room to sweep so large a circle, you can lay out your deck beams from these figures

that is, twelve or fourteen inches. Keep them down to about six inches in width, for the reason that the expansion and contraction of a very wide board will cause the seams between them to open too wide. Fasten the deck to the beams with about 1¼ or 1½-inch galvanized wire nails, with heads punched in about a quarter of an inch, and the holes filled with putty to match the wood in color, if it is to be varnished, but not necessarily so if you are going to paint it. For the sake of simplicity in construction, I have shown the forward end of the cockpit cut off V-shaped instead of a half-circle, which would require the steaming and bending of the coaming. This V shape can be cut out of straight stuff, the joints reinforced with a block of oak behind the seam, and to which each of the pieces of coaming is screwed fast from the outside. The coaming should be at least ⅜ of an inch in thickness, and one can use his own judgment as to height.

On the plans I have shown what I consider a fairly good layout of coaming and sort of cabin over the engine. This cabin consists of two flaps, hinged along the side, that can be lifted up, exposing the motor, or simply built as one hatch setting in on cleats, screwed fast to the sides of the coaming and which can be removed bodily to get at the engine. Such matters as these are of minor importance, and any man will use his ingenuity to rig it up to suit his own taste.

The placing of the engine and lining up of the shaft will, of course, have to be figured out from the engine, diameter

FIG-19

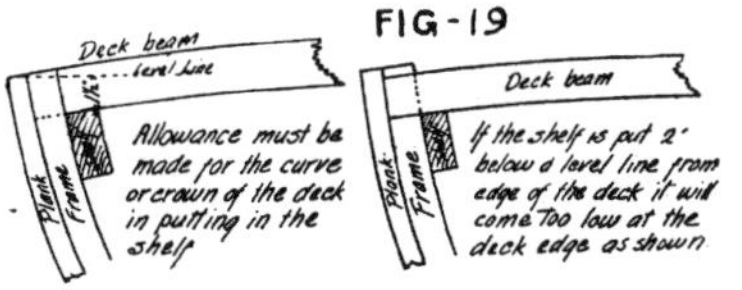

of propeller, etc., which you intend to install. It is really these points which govern the boring of the shaft hole. To determine the proper height for the engine bed, stretch a chalk line very tightly from temporary braces in such a way that it centers perfectly at both the inner and outer end of the shaft hole. By leveling across from this chalk line you can determine the height and grade for building up the beds to which your engine is to be lag screwed. On the outside of the boat, aft, this line will give you the heights for the strut, which should be put there to support the shaft, just forward of the propeller. All you need for this is the angle. Cut one end of a thin pine board so that it fits against the keel in such a manner that it forms a right angle with this shaft line and mark across it the line made by the chalk line. With this as a templet, you can make a pattern and have a strut cast, or take a piece of about ¼-inch brass about six inches wide, bend it around either a piece of iron pipe or a piece of wood, about ¾ of an inch greater in diameter than the shaft which comes with your engine. Rivet it together, and then spread the other part out in the form of two legs, which can be flanged and bolted fast to the bottom planking; by putting a shaft through this loop so formed in this sheet brass, and pouring babbitt metal around it, you can babbitt such a bearing and make one that will answer all the requirements as well as a cast strut would do.

The rudder is of the simplest construction. Take an oak board about 1¼ inches in thickness, saw out just such a rudder as you would make for any kind of a small rowboat, and attach it to the stern by two screw eyes in the transom, two in the rudder, and a rod dropped through them, forming a hinge. Fit a wooden yoke across the head of this rudder and attach tiller lines by which the boat can be steered, either by hand as a rowboat would, or if you want a more shippy arrangement, you can lead it through screw eyes or fairleaders in the deck over brass sheaves to a steering wheel which can be screwed fast to the bulkhead at the after end of the engine space.

It is scarcely necessary to go into further details on this boat, for all complications have been avoided, the main object being to tell how to build the hull in the simplest possible manner. The deck fittings and other things can be arranged to suit the owner's individual taste.

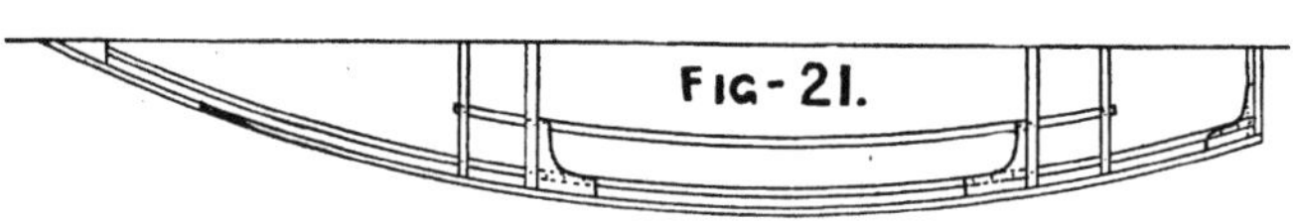

FIG-21.

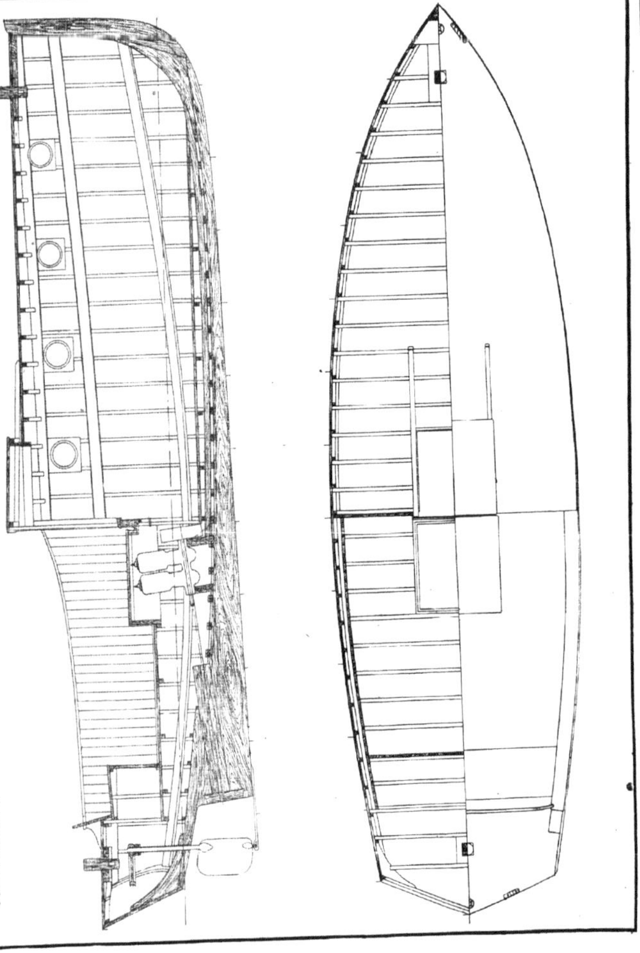

CONSTRUCTION PLAN OF 25-FOOT NOCK CABIN CRUISER

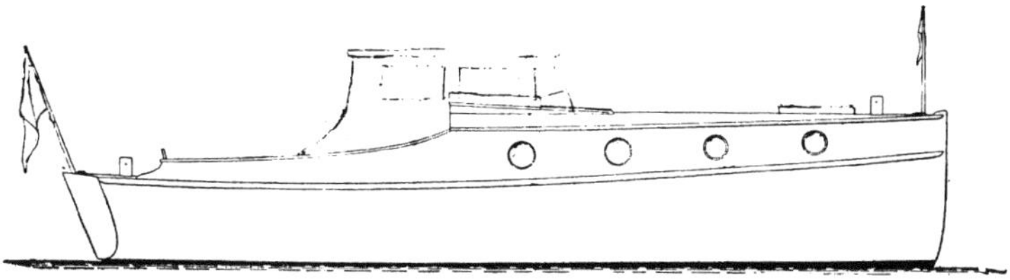

A 25-Foot Cabin Cruiser

BY FREDERIC S. NOCK

FOR such readers as will be interested in building a small cruising motorboat, I am going to try and describe how to proceed to construct a 25-foot boat, and if these specifications are carefully adhered to, those who build will be surprised to find what a handly little craft it will make.

The engine for which the boat was designed is a two-cylinder, two-stroke, rated at 6-hp. and the weight allowed for same is 350 pounds. Therefore, if you do not use one of these engines you at least should bear in mind that in order to get the best results, you should install an engine of about the same weight and power, irrespective of whether you desire a two- or a four-stroke engine. Do not install a high-speed, high-powered engine if you want a cruiser, and bear in mind that this craft is not designed for a speed boat. Eight (8) miles an hour can be depended upon with the outfit mentioned, and that is speed enough for so small a cruiser.

It is almost a moral impossibility to cover all the details in the specifications even though they are much more comprehensive than such as I should furnish to a practical builder, but I will try to explain as closely as possible how to proceed.

When you have laid down the lines full size, you can proceed to get out the moulds. For this you will need about 100 feet of hemlock, spruce or some such material. Saw to shape and *don't forget that the lines are to the outside of the plank,* therefore if you mark your moulds the same as the lines, you will have to take off ¾ of an inch around same, as these moulds are to represent the *inside* of the planking and the planking is to be ¾ of an inch in thickness. Of course you can take this amount off the outside when you lay down the lines if you prefer, and by doing so, you will save yourself considerable labor.

A practical boat builder seldom uses all the moulds, but would probably use Nos. 2, 4, 8, 16, 20, and 22, and spacing them 4 feet apart except the two end ones, you can proceed in this manner and have good results provided you use good strong battens to make a fair curve and hold the frames in place without bending the battens.

The lines as laid down usually show but one side of the mould as both sides should be alike, therefore you will find it easier to mark out one side of the mould on a piece of board, fasten this to another piece and saw the two of them out at the same time; this applies especially to the case if a band or jig saw is available. Mark and cut out all your moulds in this manner and then proceed to fasten them together. There are many different methods in vogue, but one that is about as good as any for the amateur is to put the cross pawls in such a position that the lower edge of all of them sets to the sheer line on each mould. Where more than two pieces of wood are used to make a mould, you can cleat them together, taking care to keep the cleat well in from the edges in order not to interfere with the frames, etc., if any of them should be close to the mould. Cleat the lower part of the moulds with a piece of spruce or some such material about 2x2 inches, so that you can use this to fasten the mould to the keel.

If the cross pawls are set true, it is a great help in setting up, as you can level same, and again I would call your attention to the necessity of *marking the center line on each mould.* I believe that in a case of this sort it is also a good plan to mark on the moulds the waterline, for by so doing you can easily ascertain whether the moulds are all set up correctly or not before starting to put on the battens. The practical builder could tell this at a glance, but it is not always so with the amateur.

When you have completed your moulds, you can proceed to mark out the keel. For this you will need a nice piece of oak, about 21 feet long, 3 inches thick and 12 inches wide; carefully smooth the sides and mark out on same the shape, and the stations, cut to the lines and after carefully trimming the keel run in a line through the center of the upper side, and, with your square, mark the stations across the top. Select a good piece of oak for the stem, mark the shape on same and cut to the lines. It is a good plan to make a thin wood template of the stem as you can use this template to mark out the shape and also the rabbet line. To do this, you can bore a number of small holes along the line of the rabbet, and with a drill or awl you can easily reproduce this on one side of the piece of wood you have previously cut to shape, reverse the template and mark the other side, draw a center line down the face of the stem and a line ⅜ of an inch on each side of same, which allows ¾ of an inch for the face of the stem. Trim from the rabbet line to these two outer lines and then proceed to cut the rabbet, using a fid (Fig. 3) to get the bevel. If you want to get the rabbet on the stem absolutely correct you can do so by marking on same the waterline and cutting the rabbet to the exact bevel, but I do not think you would save much time by doing this, for when the stem is in position and you start to place the battens, if the rabbet line needs fairing in places you can easily accomplish same, having the battens to work to.

If the engine you decide to use is of the dimensions as the one called for on the plans, you can rest assured that the line of the shaft will not differ, and therefore you can proceed to get out the shaft-log. For this you will need

LINES OF 25-FOOT NOCK CRUISER

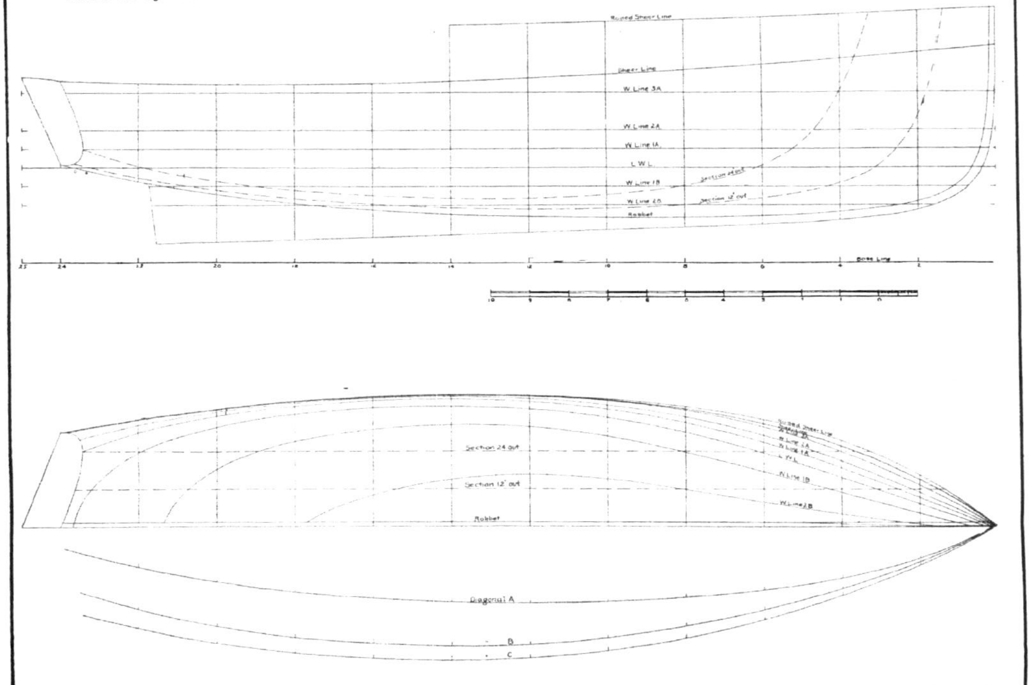

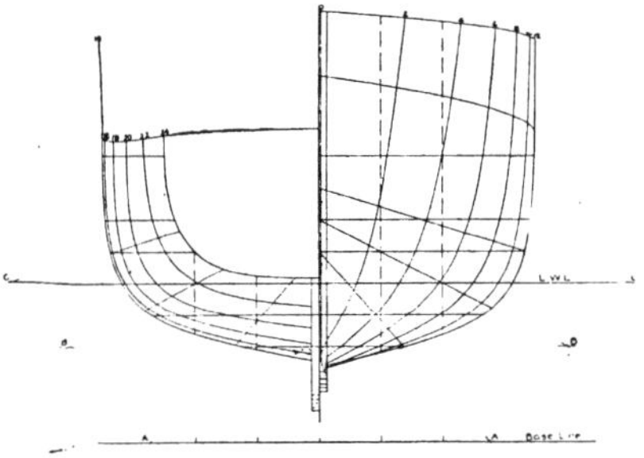

Body Plan

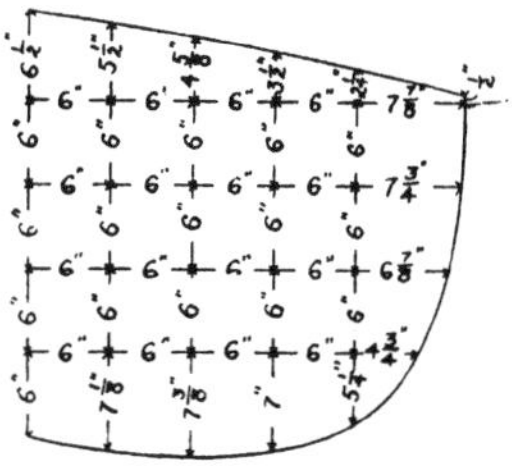

Plan of One-half Transom

LAYING DOWN TABLES FOR LINES No. 186

All Dimensions given in Feet, Inches, and Eights, and Above Base Line 2'-6" below the L.W.L.

All Lines to outside of Plank

	Stations	0	2	4	6	8	10	12	14	16	18	20	22	24	25
HEIGHTS	Raised Sheer Line	6.9.0	6.7.7	6.6.7	6.6.0	6.5.2	6.4.4	6.4.0	6.3.6						
	Sheer Line	5.9.0	5.6.6	5.4.6	5.2.7	5.1.2	4.11.6	4.10.5	4.9.5	4.9.0	4.8.6	4.8.6	4.9.1	4.9.6	
	Rabbet		1.7.1	1.3.7	1.2.5	1.2.1	1.2.0	1.2.3	1.3.2	1.4.4	1.6.4	1.9.3	2.1.5	2.7.0	
	Keel Bottom		1.4.1	1.1.2	0.11.6	0.10.7						0.6.3			
	Section 12" out		3.10.4	2.1.0	1.8.3	1.6.1	1.4.7	1.4.6	1.5.2	1.6.2	1.8.1	1.10.6	2.2.5		
	" 24" "			4.6.5	2.6.6	2.0.2	1.9.3	1.8.2	1.8.3	1.9.2	1.10.7	2.1.4	2.6.3		
HALF BREADTHS	Raised Sheer Line		1.4.7	2.3.4	2.10.1	3.2.2	3.4.5	3.5.6	3.6.0						
	Sheer Line		1.3.2	2.1.6	2.8.6	3.1.3	3.4.2	3.5.5	3.6.0	3.5.3	3.3.7	3.1.3	2.10.1	2.6.0	
	W.Line 3 Above		1.1.3	1.11.7	2.7.6	3.0.7	3.4.0	3.5.4	3.5.7	3.5.2	3.3.6	3.1.2	2.10.0		
	" 2 "		0.11.1	1.9.1	2.5.1	2.10.7	3.2.5	3.4.6	3.5.4	3.4.7	3.3.2	3.0.4	2.8.3		
	" 1 "		0.9.4	1.6.7	2.2.7	2.9.0	3.1.2	3.3.6	3.4.5	3.4.0	3.2.2	2.11.1	2.5.6		
	L.W.L.		0.7.3	1.3.7	1.11.4	2.5.7	2.10.5	3.1.3	3.2.2	3.1.5	2.11.5	2.7.4	1.11.3		
	W.Line 1 Below		0.4.3	0.11.0	1.5.4	1.11.5	2.4.6	2.7.6	2.8.3	2.6.7	2.3.0	1.6.2			
	" 2 "		0.0.7	0.4.1	0.7.7	0.11.6	1.3.1	1.4.5	1.3.0	0.10.0					
Diagonal A			0.9.7	1.4.2	1.8.1	1.10.4	1.11.7	2.0.3	2.0.0	1.10.7	1.8.7	1.5.6	1.1.0		
" B			0.11.0	1.7.5	2.2.3	2.7.4	2.11.2	3.1.3	3.1.7	3.1.0	2.11.1	2.7.6	2.2.3		
" C			1.0.0	1.9.6	2.5.4	2.11.2	3.3.2	3.5.4	3.6.2	3.5.6	3.4.0	3.1.0	2.8.3		

Diagonal A Intersects Perpendicular 3'-0" Above Base Line, and Base Line 2'-9" out

" B " " 3'-6" " " " " W.L. 2 B. 4'-0" "

" C " " 4'-0" " " " " L.W.L. 5'-0" "

two pieces of oak, 3½ feet long, 3x3 inches, plane them smooth and with your gauge mark a center line on one side of each piece, and another line ⅝ of an inch out on either side, scribe a half circle with a radius of ⅝ of an inch on either end, and proceed to cut this halfround piece away. If you are so situated that you can obtain the use of machinery, this job would be simplified by running a few saw scarfs through same to the required depths; then, with a gouge, cut away the superfluous stock, finishing it smooth with a round plane. In some cases the boat builder will run a saw scarf through the center line of both pieces, clamp them together and with a spur auger of the right size, bore the hole through, but no matter how much care is given, the spur is liable to run to one side or the other, and not follow the saw scarf, and if this happens you will spoil your shaft-log. If you happen to have an auger of the right size, it would do no harm to run it through the center of the log after you have cut the two pieces out to shape, for in this manner you can rest assured that the hole is the correct size and round. Do not try to spline this shaft-log, as there is but little stock on either side for the fastenings and by putting a stop water at either end and calking between them you can make the seam tight. The horn timber can now be gotten out, the specifications call for this to be sided 3 inches. Select a nice piece of oak, mark and cut to shape, draw a center line along the upper side and then proceed to get out the knee for the transom. This can be an oak or hackmatack knee, a natural crook is to be preferred, to be about 3 inches in thickness. Cut to shape but do not fit until you are ready to set up the keel.

You will need another knee to connect the stem to the keel, which same can be a natural crook or cut from a plank, mark it out on the plank so that the grain runs from point to point in order to get the greatest strength, it is to be 3 inches in thickness. When you have cut it to shape, carefully fit same, taking due care to see that all joints are close and when fitted to the stem and keel, the stem does not lean forward or aft, but is the same position as shown on the plans. Clamp the knee in position, bore and bolt together with ⅜ inch diameter galvanized iron, sink the heads of the bolts well into the outside of the stem and keel so as to allow for covering same with wooden plugs. The inside ends should be riveted over galvanized iron washers, or if you prefer, you can fasten the knees in position with galvanized iron screw bolts; if a washer is put under the nut, you can draw the pieces together in good shape.

When the stem is fastened to the keel you can proceed with the shaft-log. Set this in position and bore through each side about 4 inches from the forward end, and bolt through keel with 5-16 inch diameter bolts, countersinking the heads in the underside of keel and drawing up with nuts on washers on the upperside of the log.

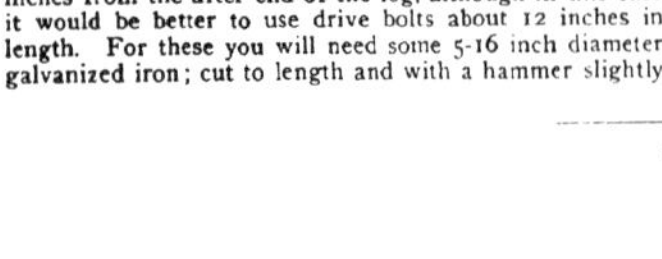

Fig. 3

Bore and put in two more bolts of the same size about 8 inches from the after end of the log, although in this case it would be better to use drive bolts about 12 inches in length. For these you will need some 5-16 inch diameter galvanized iron; cut to length and with a hammer slightly taper one end. Then fasten in a vise and hammer a head on the other end. Bore through the shaft-log with a 5-16 inch diameter bit, and into the keel with a ¼-inch diameter bit or auger. Set the horn timber in position and, after carefully fitting same, secure it by clamps or some other convenient method, and proceed to fasten it; if you start at the forward end it would be well to bore

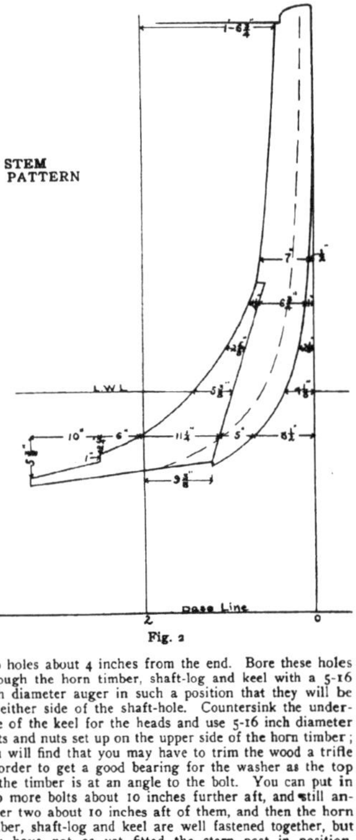

Fig. 2

two holes about 4 inches from the end. Bore these holes through the horn timber, shaft-log and keel with a 5-16 inch diameter auger in such a position that they will be on either side of the shaft-hole. Countersink the underside of the keel for the heads and use 5-16 inch diameter bolts and nuts set up on the upper side of the horn timber; you will find that you may have to trim the wood a trifle in order to get a good bearing for the washer as the top of the timber is at an angle to the bolt. You can put in two more bolts about 10 inches further aft, and still another two about 10 inches aft of them, and then the horn timber, shaft-log and keel are well fastened together, but you have not as yet fitted the stern post in position, although the keel is cut for it.

Select a piece of oak that will work out 3 inches square and about 15 inches in length, and cut to fit the space allowed for same. When you are satisfied that it fits properly you can fasten it in position with 5-16-inch drive bolts.

The transom, which the specifications state is to be of oak, is 1¼ inches thick. You will need two pieces 30x36

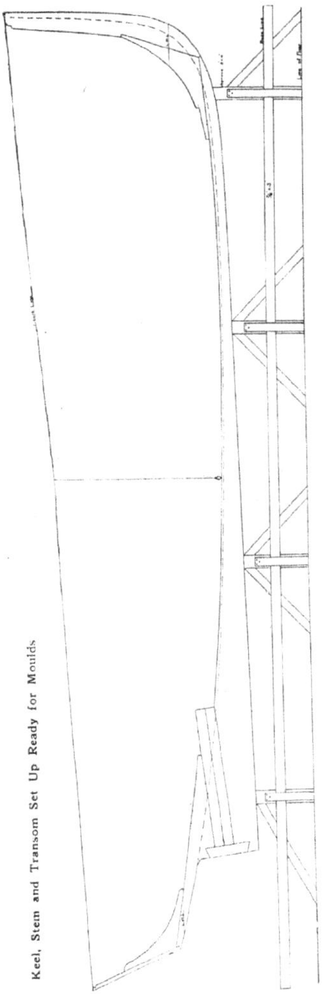

Keel, Stem and Transom Set Up Ready for Moulds

inches, and inasmuch as it might be a difficult matter to procure oak 30 inches wide, you will probably have to use two narrower pieces and bolt them together, or, if you prefer, you can use mahogany for the transom. There is little or no difficulty in obtaining this material 30 inches wide, but it would cost a trifle more than oak. The plans show one half the transom with all dimensions marked on it. You can set this out direct on one of the pieces of stock for transom if you desire, and saw it out and mark the opposite side for same. Or you can saw the two pieces at one time with a band saw. Trim carefully to shape and get out a couple of cheek pieces of oak 1 inch thick and about 3 inches wide; set these almost out to the edges of the transom and fasten with galvanized screws. these pieces are to be trimmed to form a back rabbet to fasten the ends of the planking to, and you will probably have to fair them to some extent when you start to put on the battens.

You can now proceed to set up the frame. Draw a chalk line on the floor where you intend to set up the boat and to the edge of this line set up 4 pieces of spruce 3x4 inches, each piece to be one foot longer than the height on the table of offsets of *base line to keel bottom, at stations* 2, 8, 14, *and* 20. Nail a strip of wood to the sides of these upright, so that the upper edge will represent the *base line*, and if the floor is level, it will be one foot above same. Set this strip of wood level, using a long spirit level to ascertain that neither end is high. The forward side of these posts will represent the stations Nos. 2, 8, 14, and 20, and you can measure up on from the *upper edge* of the strip representing the *base line* and see that the dimensions correspond to the heights of keel bottom above the *base line* on table of offsets. These uprights must be well braced forward and aft and also at the sides in order to keep them upright and sustain the weight of the boat. It is not necessary to obtain any special size stock for these braces, as most any stock from 1 to 2 inches thick will serve the purpose if properly fastened. Set the keel on the uprights and ascertain that it is in the right position; measure from the *base line* to the *waterline* on stem and see that it is 2½ feet above the *base line* as per plans. Plumb the stem and securely brace it from overhead if possible. If you are sure that it is correct, you can fasten some wooden cleats to each side of the uprights and also fasten to the keel.

The knee and transom can be fitted and fastened, using plenty of fastenings but taking due care not to get any nails or screws where they will interfere with the rudder port, which will have to be bored and time spent looking out for this will mean a saving in the end, as nails and screws are apt to spoil an auger or bit. Place your level against the side of the knee and when vertical, stay the transom either to the rafters or some other convenient place. Then take your chalk line and stretch it from the center line of the transom to the center line of the stem, haul taut, and drop a plumb-bob from same (see Fig. 8). If the point of the bob touches the line on the center of the keel you can rest assured that the stem and stern are in line, but do not be satisfied with one trial, try the plumb-bob at different points and if any of them are out, sight along the side of the keel and if this is not in a straight line fore and aft you must brace it until it is perfectly straight and the plumb-bob must intersect the center line at any point. When you have it correct, you can fasten the stays or braces strongly, so as to keep it in position.

"Consort," Built from Mr. Nock's Plans and Instructions

PART II.

BEFORE you begin to set up your moulds, it would be advisable to finish the *rabbet line,* as you have only cut the stem. The *rabbet line* is supposed to be marked on the keel on either side, and inasmuch as the upper side of the keel as far aft as the shaft-log represents the inside of the planking, it is not a difficult matter to trim this rabbet at intervals, taking the proper bevel at the different stations from the lines you have laid down. Then fair up the places between the points you have spotted, or cut to the proper shape. Where the rabbet line crosses the knee and connects with the rabbet line on the stem you must be very careful to get it correct.

It would be advisable to get two pieces of oak, 1 inch thick, about 6 feet in length, to fasten to each side of the shaft-log and horn timber; they should be cut to the curve of the rabbet line and the proper bevel, and securely fastened. Cut out places in the lower edges of these pieces to receive the heels of the frames; of course you will understand that these pieces are to form a back rabbet for the garboards or the plank nearest to the keel and should be cut to the proper bevel of the inside of the plank.

Having finished cutting the rabbet line along the keel, etc., you can now proceed to set up the moulds. Starting with No. 2, set this in position so that the forward side of the mould intersects with station No. 2 on the knee. Cut it over the knee and set it down so that the lower part of the mould intersects with the back rabbet; measure up from your *base line* to the L. W. L. on the mould and if it is 2½ feet up, screw through the block into the knee to hold the mould, plumb the face of the mould and when you have leveled the cross pawl you can secure the mould with stay-laths. Moulds 4, 8, and 12 can be set up in the same manner with the forward sides of the moulds intersecting with stations 4, 8, and 12 on the keel, but the moulds 16, 20, and 22 are to be set with the after side of the moulds intersecting with their respective stations. Perhaps it might be well to explain the reason for setting them in this manner and I would call your attention to the fact that the edges of the moulds are cut at right angles to the face and the shape of curve of the boat commences to narrow up from station 12 forward and from station 16 aft; therefore these moulds are so placed that the edges which the battens spring against are in a direct line with the stations for which the moulds were made. When all the moulds are properly stayed, it is a good plan to go carefully over them and see that they are perfectly plumb and the cross pawls level, and if you stretch your chalk line fore and aft in a line with the L. W. L. at the stem and stern, it should also intersect with the waterline on each mould. If it does you can rest assured that your moulds are set up properly. Fasten some stays from one mould to another and start to bend and fasten on the battens. For these you will need some nice, clear pieces of yellow pine or fir about 1¼x1½ inches; if you get them long enough to extend from end to end in one piece so much the better, otherwise you can use shorter lengths and let them lap by one another. Fasten the end of one of the battens in the rabbet on stem, bend carefully around the moulds and fasten to same with long thin screws, putting washers under the heads of same to prevent them pulling through the battens. Set the first batten so that the lower edge touches the mark representing the sheer line on each mould; when you get to the transom you will probably find that the bevel of the cheek pieces does not allow the battens to bear the entire width, and in that case it must be pared until the batten has a good bearing, and the same thing applies to the other battens where they are fastened to the transom and stem; you must also see that there is the right bevel cut on the edge of the transom. It would be advisable to put about six battens on a side, space them about the same distance apart. If you have been careful in making and setting the moulds, you will find that the battens touch each one and at the same time have a nice fair curve.

The frames, or timbers, should be cut from straight grained white oak; if you can procure some stock that has not been dried you will find it much easier to bend when it has been steamed than the seasoned stock. Under no conditions should you try to use kiln dried stock; it is almost an impossibility to bend it, and even if you do succeed in bending it you will find that it has a natural tendency to straighten instead of retaining its shape. The specifications call for the frames to be 1⅛x1⅛ inches, spaced 9 inches center to center. You will need 30 frames for each side. If, however, the frames from station No. 4 to the forward end of the shaft-log are in one piece from sheer to sheer, it will reduce the number of frames. You will need nineteen frames to run from sheer to sheer, and as you are liable to break some of them, it would be advisable to get out twenty-six. The longest frame will be about 14 feet, but you had better get 16-foot lengths; there are twenty-two short frames required, and if you are using 16-foot stock you can cut the pieces in two and use for the short frames. Don't forget that you may break some of these and that it is a good plan to get out a number of extra frames; even though you do not break a great number they will not be wasted as they can be used for the floors.

Mark off on the keel the spacing of the frames, taking **care not to** measure along the top of the keel but in a **straight line,** and it is a good plan to mark the position **of the frames** on the center and upper batten as this will **save considerable** time when you are bending in the **frames.** Put some of the frames in your steam box and **let them stay** until they are well saturated, and are soft **and pliable.** It will probably take from 20 to 40 minutes, **according to** the amount of steam you have, and bear in **mind that** you do not require dry steam, it cannot be too **wet.** Having satisfied yourself that the frames are sufficiently well steamed you can proceed to bend the frames **to the battens.** If you have put in some of the long frames you can start anywhere along the widest part of the boat, gradually spring the frame down until it touches **the keel**; then, bearing your weight on same, work upwards on either side, fasten the frame to the keel with galvanized iron boat nails and clamp it to the battens, taking due care to see that it fits closely to the battens and stands perpendicular. There will be no trouble about it being perpendicular if you bring the edge to the mark on the battens. There is no need for me to tell you that this part of the work would be a great deal easier if you can have someone to assist in bending and fastening the clamps; it will also save you considerable climbing up and down, for the best way to bend these frames is to get inside of the framework and bear your whole weight on them. If you have only a few clamps you will probably need them for the next frame, and if so you can remove the clamps one at a time, and fasten the frame to the battens with a wire nail through each batten. By the time you had bent in a few of the amidship frames—which are the easiest to bend in—you will have grasped the situation and find that there is nothing very difficult about this part of the work. The frames at the forward end will require a little more care as they must be cut to fit the sides of the stem and as they do not set square with the battens, you will have to twist them so that they fit close to each side of the battens. The frames aft of the forward end of the shaft-log will have the heels fitted in the mortises in the cheek pieces and fastened to same. These frames will also have to be twisted to get the right shape to make them set close to the battens with width of the frame, and a large monkey wrench would be of valuable assistance in twisting the frames. When you have finished framing you can put in the floors. These are to be 1⅛x1⅛ inches and should be about 3 feet in length. Beginning at station No. 4, ***all the floors should be on the after side of the frames until you reach station No. 14, and from this point aft all floors should be forward of the frames.*** Bend in the floors and fasten to the keel, then put three fastenings through the side of floor and frame on either side of the keel. For this purpose it is a good plan to use a galvanized wire nail, if the fastenings are to be of galvanized iron as specified for this boat. The floors aft of the shaft-log are not bent to shape, but are to be cut out of 1⅛-inch oak; cut them so that they fit over the upper side of the horn timber and the shaft-log and shape the underside the same as the frame; fasten to the horn timber and shaft-log and also through the sides of the frames. It is a good plan to cut a piece out of the corner of these cut floors or else bore holes through same so as to allow any water that may run in aft to find its way to the lowest point, and now that I have brought up the subject of limburs, I will call your attention to the fact that there has been no provision made for same in this boat, the frames and floors are small and to cut a limbur in them that would be of a suitable size would materially weaken both the floors and frames and I would suggest that you do not cut in any limburs, but when the boat is planked, take some Portland cement and pour in each bay until flush with the top of the lowest frame and in this manner you will have a clear passage for the water and no obstructions. This is an advantage in a

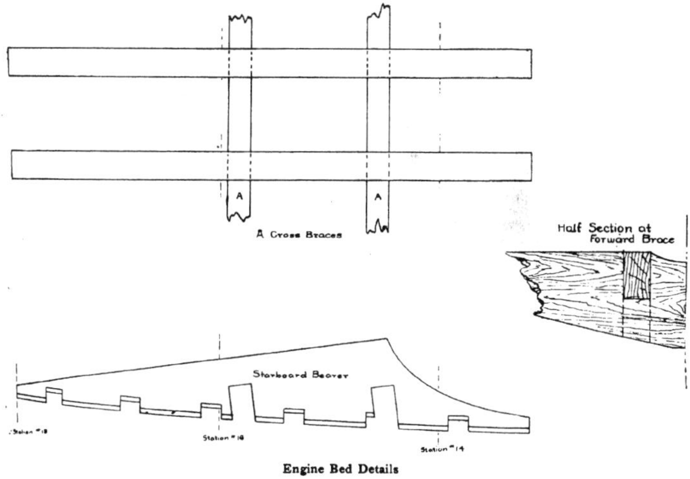

Engine Bed Details

motorboat, for it enables one not only to clean it out easily, but also when the craft is laid up in the Fall, it is much easier to remove the surplus oil and grease that will collect in the bottom of a boat of this description, no matter how careful one may be. If, however, your mind is set on having limburs cut in the frames, this is the time to do it, before you start to put on any plank, etc.

When you have completed the frames you can proceed to get out the planking. Fig. 9 shows a half section of the boat at station No. 12 and you will note that I have shown thirteen planks on each side from the garboard to the sheerstrake. It is not necessary to use exactly this number of planks to a side; some builders prefer wider planks and of course there would not be as many if they were wider. However, it is better to use the narrow plank. The *garboard*, or plank next to the keel, is usually the widest and the planks narrow from that point to the turn of the bilge upwards where they should all be of one width with the exception of the sheerstrake, which is usually made wider than the next four or five planks below it. The specifications call for yellow pine planking and if you use this material, there is no reason why you should not have it in lengths of about 28 feet, or long enough to reach from end to end without any butts. If you are unable to obtain the requisite lengths, the planks can be butted, and again if you cannot procure yellow pine or fir, you could use cedar, cypress or white pine. Either of these materials would make good planking.

As it is easier to put in the engine bed before the lower planks are put on, I am going to suggest that you start to put on the sheerstrake first; select your plank and plane to ¾ inch thickness, and before you can proceed to mark the shape of the sheerstrake, you will have to find out the shape of this plank, and in order to do so, it is necessary to take a spiling. To do this take a piece of pine or cedar about 3-16 or ¼ inch thick, 6 to 7 inches wide and about 26 feet in length, this is called a *spiling batten or staff*, and can be made up of two pieces fastened together if these are easier to obtain than one length. Bend this naturally around the moulds below the marks representing the sheerline, and with some small wire brads secure it in position, taking due care that the edge of the staff does not cover any of the marks representing the sheerline; take your pencil compasses and set them to such a width that the radius is slightly in excess of the widest place between the edge of the staff and sheerline, place the point of the compasses at the sheerline on each mould and sweep in an arc on the staff, mark the end of the staff where it intersects with the rabbet in stem, remove the staff and fasten it to the plank you intend to use for one of the sheerstrakes, and with your compasses draw an arc, the point of the compasses being set on one side of the arc on the staff. Then put the point of the compasses on the other side of the arc on staff and draw another arc on the plan. This arc will cross the first one drawn, and where these two arcs intersect or cross is the point you must work to when you draw the line representing the upper side of the sheerstrake. You must re-

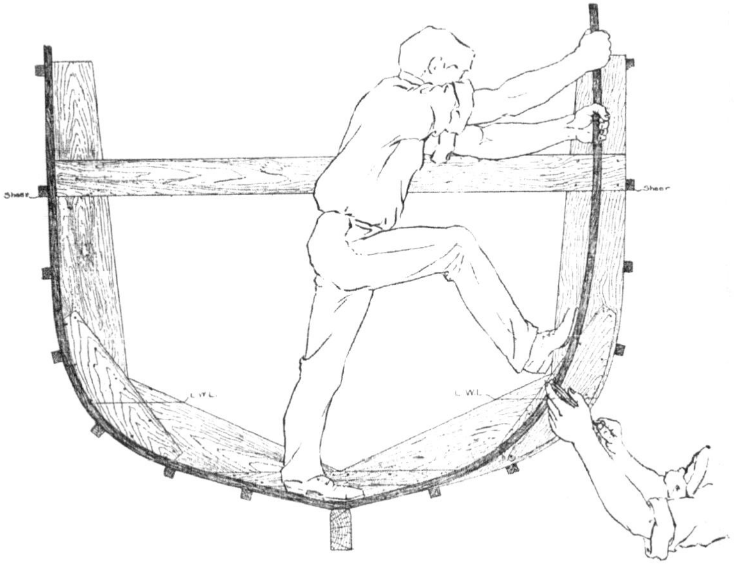

Method of Bending Frames

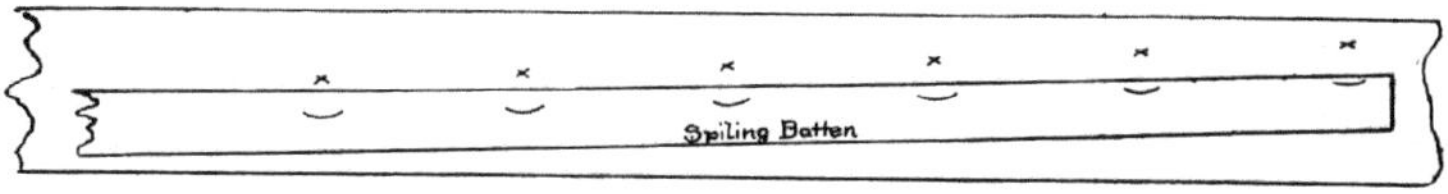

Fig. 11.—How Spots Are Transferred to Plank

peat this process at each place where you have drawn an arc on the staff, and when you have them all marked, remove the staff, select a good batten and bend to the required shape so that a line drawn along the edge of same will cut through each of the arcs you have drawn where they intersect. Fig. 10 will help to explain this a trifle more clearly, and Fig. 11 shows the staff fastened to the plank and the arcs marked on same and also the plank. To obtain the shape of the lower edge of the sheerstrake you can mark at the center the extreme width, which I have figured at 5 inches, then mark the width at each end, bend a good strong batten to these points until you have a fair line and cut to same. Bear in mind that you need two of these pieces and therefore it is advisable to saw out another one for the opposite side before you begin to fasten it to the frames. The sheerline as shown on the plans represents the height to the top of the deck and though this is short, it should be taken into consideration. The deck plank is to be ⅞ of an inch in thickness, therefore the sheerstrake should be ⅞ of an inch below the sheerline marked on the moulds. Fit the forward end to the rabbet on the stem and fasten it, taking due care that the upper edge touches the marks ⅞ of an inch below the sheerline; hold it in position with clamps and fasten to the frames, countersinking for the nail heads so as to allow of their being covered with 7-16 inch diameter wood plugs. The after end must be carefully cut to fit the bevel of the transom and be well fastened to the cheek pieces on same. It is a good plan for the amateur to mark on the edges of the moulds or frames the width of the planks. Fig. 9 shows a half section at station No. 12 and you will note that there are thirteen planks to a side including the sheerstrake and garboard; the upper strakes, representing the raised sheer, I have not taken into consideration as you are working from the sheerline. Starting from the sheerline on the mould representing station No. 12, make a mark ⅞ of an inch below same, then another one 5 inches below that. This will represent the sheerstrake. Now start at the keel and mark 8 inches up, from this mark 7 inches upwards, and another 6 inches, and the remaining spaces you can divide equally into nine spaces. These will represent the greatest widths of the planks, and you should proceed to mark on the stem the height of the upper edge of the garboard and the lower edge of the sheerstrake which should be considerably narrower at this end than at station No. 12. The transom should be marked in a similar manner and then, if you desire, you can mark out on each mould the widths of the planks; of course the garboard and the next two strakes will be wider in proportion than the others, but it is a simple matter to figure this out and have a good curve to all the planks.

The professional builder does not always lay out his plank in this manner, and some of you may know a much better method; this article, however, is not written for the practical boat builder, but for the amateur. Inasmuch as you are not going to fit in the garboards at present on account of the engine bed, we can at least work to the width laid out for the garboards and other plank and proceed to get out the next plank below the sheerstrake. You can take a spiling from the lower edge of the sheerstrake in a similar manner to the spiling of the sheerstrake, except that in this case you work from the lower edge of the sheerstrake in sweeping in the arcs on the staff. Be sure to make some marks on the sheerstrake and staff so that when you have cut out your plank you can bring them to the same position as you had when you took the spiling. Mark the widths of this plank at intervals, bend a batten to it, draw a fair curve through these points and you have a line representing the lower edge of this plank. Cut out two planks to this shape. If you take this plank and bend around the moulds in order to see that it fits, you will most probably find that the seam is open quite wide in the center of the boat but it closes at the stem and the

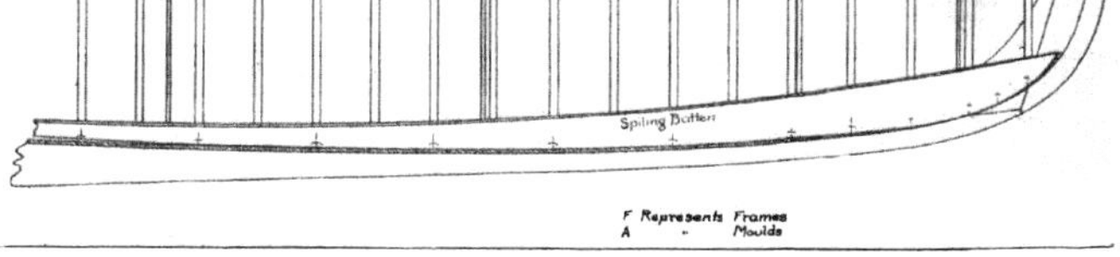

Fig. 10.—Shape of Spiling for Plank and Garboard

transom and in order to overcome this trouble you must obtain the right bevel at certain intervals; for instance, at the stem, moulds 4, 8, 16, 20, and the transom, cut the edge of the plank at these points so that the bevel at each place corresponds to the angle taken, then fair up the intervening spaces and plane the edge. This plank when fitted in position should be tight on the inside of the seam,

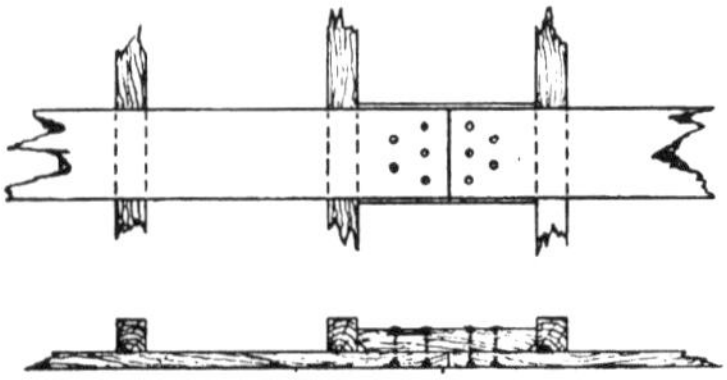

and open about 1-16 of an inch on the outside. It is really necessary to take great care to have the inside of the planks tight for if they are open on the inside the calking cotton is liable to be forced through, thus giving the water a chance to get through, and you will have a leak that will be difficult to stop. The practical boat-builder takes his bevel square, which is a small one, and, obtaining the different angles at certain points, transfers these to a piece of wood he has for this purpose, and when trimming the edge of the plank he takes care to see that it is cut to the correct angles to suit the shape. This part of the work will appear more clearly to you as you progress with the planking. When you have the plank ready to put on, fit the forward end to the stem and after clamping same securely fasten it. Clamp to the frames at intervals, and in order to get it tight against the sheerstrake, you will have to draw it up with clamps, but a better method is illustrated by Fig. 13. These planking dogs can be purchased at a small expense and it would pay to use them; however, Fig. 12 illustrates another method of wedging up the planking which is often used. Another article that is of great assistance in planking a boat and for drawing the framework together where the distance is too great to admit the use of the ordinary iron clamp is the chain clamp (Fig. 14). When you have worked out and fastened on about four planks on each side below the sheerstrake you can remove the balance of the battens, and the moulds, but before removing the moulds you should fasten the heads of the frames at intervals of about every three or four feet with stays to prevent the boat from spreading or getting out of shape. This performance is not necessary at this period unless you decide to put in the engine bed before putting on the lower planks, and the fact that you have five strakes fastened on each side should properly bind the framework, but you cannot be any too careful in putting the boat together and it is always best to prevent the boat from going out of shape than to try and bring it back after it has spread or twisted. Fasten the ends of some of the frames to the rafters with staylaths and after you have removed the moulds, ascertain that the boat has not changed shape and is level, and you can proceed to put in the engine bed.

If you intend using the same engine as mentioned in the specifications, or have on hand the dimensions of the engine you intend using, you can get out the two fore and aft bearers. For these you will need two pieces of oak 3 inches in thickness, about 5 feet long and 11 inches wide. The center line of the shaft will be directly over the center line of the keel, and as you have already bored the shaft alley, your line of shaft must correspond with it. Fasten a piece of wood across the hole in the stern post, cut a notch in same that will be directly in the center of the hole, fasten a piece of very strong fine fish line or piano wire around same, and lead the other end through the shaft alley. Carry it forward about up to station No. 12; fasten a piece of wood across the frames at such a height that the line when drawn tight over the top of same is directly in line with the joint in the side of the shaft-log. If it does not come out correctly the first time, you will have to raise or lower the forward end until you are sure that it is quite correct, and also in line with the center line of keel. The line will represent the center line of the shaft and your measurements can be taken from it. First ascertain the length of the base of the engine where it should rest on the fore and aft bearers, then find out the distance from the center of the crank shaft to the underside of the base of the engine, and also the width of the base below the flanges, for this latter dimension will determine how far apart you are to set the fore and aft bearers. Measure out on each side of the center line on the keel half the width you require for the inside of the bearers, and make a mark on the frames, take a piece of thin wood about 5 feet in length, shape and set it to the line marked on the frames, use your level to ascertain that it is plumb, and fasten it

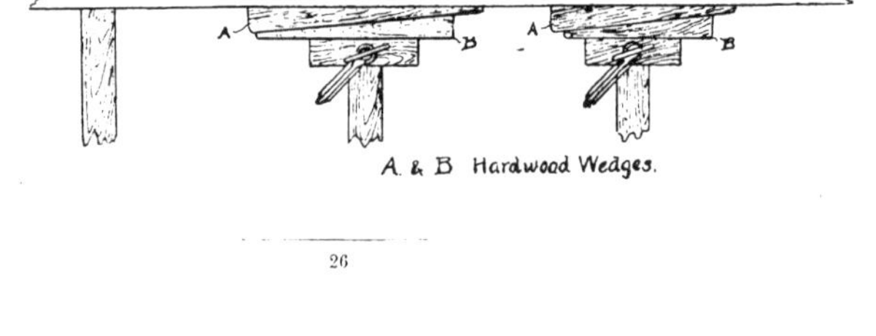

A. & B Hardwood Wedges.

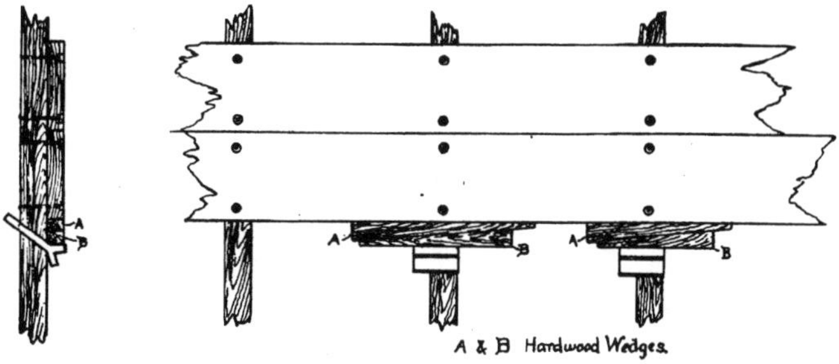

Fig. 13

temporarily. Now take your level and bring it up so that it just touches the under side of the line, see that it is horizontal and make a mark on one end of this template; repeat the operation at the opposite end, draw a line through the point, and you have the line of the center of the shaft marked on same.

You can now proceed to get out the two bearers, adding to the size marked on the template the thickness of the frames, or better still, a quarter of an inch in excess of same, to allow for trimming, or, if you prefer, you can get out a template to fit exactly over the frames and floors and mark the shape of these bearers from same. It is simply a matter of choice which method you use. If you cut the underside of the bearers to the shape of the template, set in position, mark both sides with the compasses in order to get the height of the frames on the inside and outside, then mark the frames and floors, remove the bearer, cut to shape, and see that it fits properly; repeat this operation on the piece of stock for the other bearer, set in position, clamp it so as to keep it from moving and ascertain that it fits properly and is in line both fore and aft and also in a vertical position. If there is any part of these bearers that extend below the under side of the frames, you can easily remove the superfluous stock when you have the bearers fastened in, before you start to plank. The cross logs are 2½ inches thick, and for these you will need one piece of oak 6 feet long and 1 foot wide, and another piece 5 feet long and about 8 inches wide; set these pieces on the keel and the places indicated on the plans, or to suit the base of your engine if it is a different make or style, fasten them temporarily and mark out the shape of the underside of same on the forward and after sides; if you do this properly you can cut to the correct bevel the first time, and they will fit close to the inside of the plank where this part of the boat is planked. Cut the forward pieces to a depth of 5 inches to allow the fore and aft bearers to fit into same for that depth, fasten them securely to the keel, and then put in the fore and aft bearers, cutting them over the cross logs, securely bolt to same, and fasten up through the under side of the frames into the bearers. If the under side of the flange of the engine bed is below the center line of the shaft a certain distance, the bearers will have to be cut away that amount in order to have the shaft in line, and if the under side of the flange is above the center line of the shaft, the bearers will have to be raised; this should be ascertained before the upper sides of the bearers are finished.

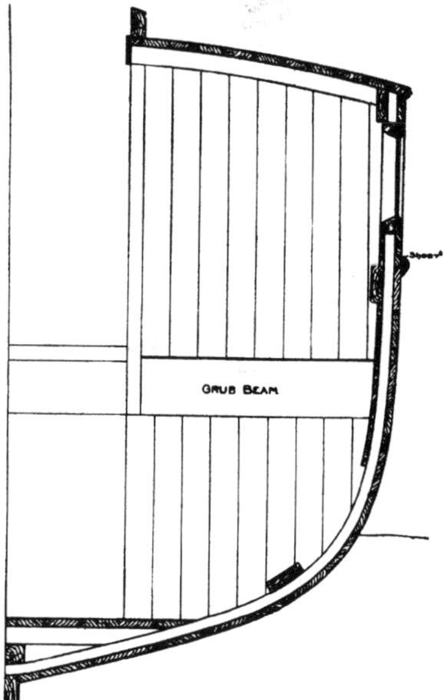

Fig. 9

Another View of "Consort"

PART III (*Conclusion*)

The keelson can now be gotten out and fastened in position. For this you will require a piece of oak, or yellow pine, 2 x 4 inches, 12 feet in length. Trim the forward end so that it fits the scarf in the stem, and bend to the frames. If any of the frames at the forward end have a tendency to keep the keelson from bearing tight on the frames and floors directly over the keel, you will have to trim the keelson and when you are satisfied that it fits properly, proceed to fasten it, using through fastenings at the forward end and at each alternate floor.

The bilge clamps can be fitted and fastened to the frames before the planking is fitted, and it would be as well to do this part next. The specifications call for the bilge clamps to be of yellow pine 1¼ x 4 inches amidships, tapered to 1¼ x 3½ inches at the ends. This taper should be about 7 feet in length. Select some straight grained piece of stock and have it milled to the required dimensions, and finish with a hand smooth. Mark on the center frames the location of the bilge or clamp on either side, also the stem and stern, bend the bilge clamp down in the center and fasten with a screw clamp and then work the ends to the required position, or as a good bearing on the frames, bore and fasten to the frames, putting one fastening through each frame at the upper and lower edges of the clamp. These fastenings should be either riveted or clinched over the frames. Having fastened in the two bilge clamps, you can continue your planking and I would suggest getting out the garboards first. You will proceed in a manner somewhat similar to getting out the sheerstrake, but owing to the fact that this is more difficult, I will explain it so that you can proceed without any trouble.

Take a piece of pine or cedar about the same dimensions as your spiling batten or staff, but don't use that one as you will need it for the balance of the planking; tack this over the frames close to the rabbet in the keel, take your compasses, and setting them to a radius of about one-quarter of an inch more than the widest place between the rabbet and the edge of the staff, strike in a number of arcs at intervals of about one foot apart until you get to the short turn at the stem, where the marks should be quite close in order to get the shape more accurately. Make some marks on the keel and batten in order to enable you to get the batten back in the same position again. Fig. 10 shows the staff in position with marks showing the spiling, and if these marks are transferred to the piece of plank to be used for the garboards in the same manner as suggested for the sheerstrake, there will be no difficulty in making them fit, but it may save you some lumber as well as considerable labor if you first cut the spiling batten to the mark you have made, pare and trim until it fits the rabbet perfectly, and then it is no trouble to mark around same with a pencil, and you have the shape of the lower edge of one of the garboards. The shape of the other edge is obtained in the same manner as the lower edge of the sheerstrake, mark the greatest width, the width at the forward and after ends, and a couple of intervening places if you so desire, bend a batten to same, strike a fair curve and cut to shape.

You will have to exercise considerable care in fitting the lower edge to the rabbet, chalk the inside of the rabbet, and when you have clamped the garboard in position, if there are any places where it needs trimming to make a good fit, don't slight it, but spend time enough to have it fit all along the rabbet. Mark on a board a duplicate of the plank, cut and fit as carefully as the first one and, when they are ready, steam them well and start to fasten one of them, beginning at the stem. Drive the plank tight into the rabbet, bore a hole ⅜ or 7/16 inch diameter, to take the wood plug to cover the head of the fastening. This should not be over ¼ inch deep; then bore a hole a trifle smaller than the nail through the planking into the stem, and fasten with galvanized boat nails. The fastenings should be about 1¾ inches apart where the garboard connects with the stem and in the keel they should be about 3½ inches apart. There should be three fastenings in the width of the plank at each frame and floor, two of the fastenings to be through the frame and one through the floor, unless you make the after end very narrow, and in that case you would not need more than two fastenings at each frame and floor.

When you have finished fastening the garboards, you can proceed to get out the next plank; cut to shape and fasten to the frames, etc. Before you continue to plank any further, I must call your attention to the planks which are to be used around the turn of the bilge. These planks should be cut out of stock that is thick enough to allow of the inside being coped or planed hollow to fit close to the frames. When you have removed sufficient of the inside of the plank to admit of its fitting closely to the frames, set your gauge to ¾ inch and run a score along the edges of the plank, then plane to this line, but not for the entire width of the plank, as you can easily plane this when you finish the outside of the planking.

You can now cut out and fasten in position the next five strakes, and when you have finished with them you will be ready for the shutter, for such is the name given to the last plank to be fitted. It is necessary to take a spiling on both sides for this plank, and be sure to get it large enough. Having satisfied yourself that it is the required shape, fit the forward end to the rabbet in stem and drive it in place, fasten to the stem and continue to drive it in till it fits close to the frames and fasten to same as fast as you drive it in place, for in this manner you will prevent it from springing out when driving. The

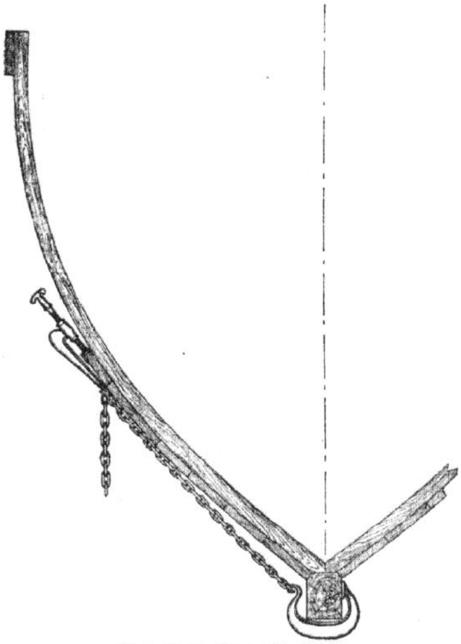

How Chain Clamp Is Used

cutting of the after end of the plank to fit the rabbet in the transom must be done before the fastenings are put in the last few frames; finish fastening in frames and transom and fit the shutter on the opposite side. A well-fitting shutter tends to tighten up all the planking, and if you have kept the widths of the planks about equal, it is hard to distinguish the shutter from the other planks.

The planking of the raised sheer can now be gotten out and fastened in position, and as I have already explained how to obtain the shape of the planks, there is no need for me to explain the operation as there are but three planks to be gotten out for either side of the raised sheer. I have figured that all the planking you will use will be full length, but for the benefit of those who are unable to procure the long lengths or prefer the shorter planks, I would state that the method of planking would be the same, the only difference is that some of the planks would be in two or more pieces. All that is necessary is to make the butts come between the frames and fasten them securely. Fig. 16 shows the usual method of fastening a butt. The butt block should be of oak about 1 inch thick and of sufficient width to lap over the edge of the plank on either side ⅝ of an inch. The forward and after ends of these blocks should be chamfered on the side nearest to the inside of the plank so as to allow any water that might leak in to run through, instead of collecting on top of the block.

Fit all blocks tight between the frames, and if properly fastened, the plank is stronger at this point than elsewhere. If you have butts in the planking, bear in mind the necessity for distributing same. Do not make one butt come directly over the top of another, but break all joints by placing at least three planks between, if the butt is between the same frames.

When you have finished planking the boat, take a jack plane and roughly plane the outside, then the boat is ready for calking. It is much better to have this done by a man who makes a business of it, as there is a great deal more in calking a boat than merely driving in cotton. If you desire to try your hand at this, you should procure some good spun cotton. You will also need some calking irons and a mallet—one of these irons will be needed for opening the seams and the other for driving in the cotton. Look carefully over the seam you intend to calk, and if you decide that it is open enough to drive in the cotton, select a piece and start it in with the iron, using the calking mallet to strike the iron. Do not stretch out the cotton and drive it in, but bring it back a trifle, then drive; keep on doing this—it makes a lap and thus makes a much thicker bunch to drive in. Set it down hard and keep it about ¼ of an inch below the surface. Experience will help you more than any explanation I can give you, and you will find as you progress just how much cotton to put in the seams, and how hard to drive it. If you want to find out just how easy, or how difficult, it is to calk a boat, just watch some of the men calking a small yacht or vessel, then go back and try it. When you have the hull calked, the seams should be painted with a thick lead paint; a long narrow brush with short bristles, known as a seaming brush, is the best thing to use to run the paint in the seams. This painting of the seams makes them hold the putty and also form a ridge which prevents the cotton from working out, as it would have a tendency to do when there is much jarring on the boat such as would be occasioned by riveting in the deck clamps, etc.

The deck clamps and raised deck clamps can now be gotten out and fastened in position. The clamp is to be of yellow pine, 1¼ x 4 inches, the upper and lower edges to be chamfered from the forward end to the bulkhead at the after end of the cabin. Set the upper edge of this clamp the same height as shown on the construction plan and fasten to the frames. It is a good plan to fasten through the sheerstrake, frame and clamp, at each frame, setting these through fastenings alternately at the upper and lower edges. The raised deck clamp is to be of yellow pine 1⅛ x 3½ inches. Chamfer the lower edge and fasten

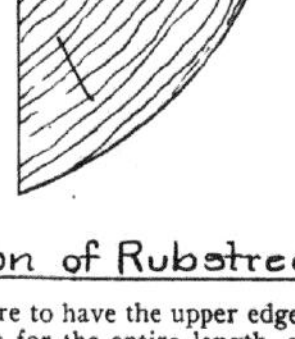

Section of Rubstreak

in position. Take care to have the upper edge of the clamp at the correct height for the entire length, and when you have put in a few fastenings it is advisable to ascertain that the upper edges are true, and the same height on

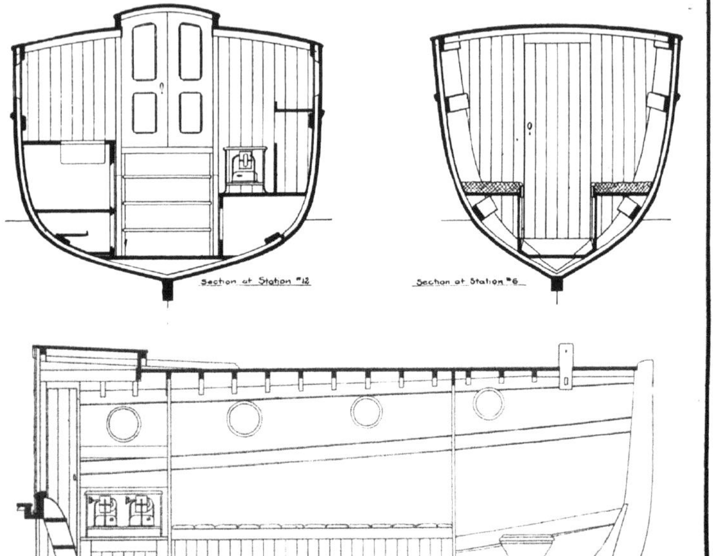

PROPOSED CABIN PLAN FOR THE NOCK 25 FT. CABIN CRUISER

both sides of the hull, and if so, you can finish fastening same, adopting the same method of through fastening as suggested for the clamp at the normal sheer.

If you intend to finish the exterior of the hull before doing any work on the interior, you can proceed to get out the two knees for the stern; these are to be cut from oak plank 1¾ inches in thickness. Fit a piece of oak 1¼x1¾ inches across the inside of the transom at such a height as to catch the ends of the deck plank, and then fasten the knees in position, securing same to the clamp, sheerstrake and transom.

You will note that there are but three deck beams shown on the plan; these are to be of oak, sided 1¼ inches, moulded 1¾ inches, and cut with a crown of 1½ inches in 5 feet. Mortise the ends of the two forward frames into the clamp and fasten to same. The after beam is to be halved into the knees and fastened. Between the two after beams you can fasten a piece of oak 8 inches wide and 1¾ inches thick to receive the lower end of the towing post or bitt, and then proceed to plank the deck. You will need some nice, clear white pine for this plank, and unless you can obtain the pine already cut with the edge of the grain on the flat side of the plank it will be best to purchase 2-inch plank and have it sawed and planed to the required thickness. The edges should be planed so as to have a bevel of 1-16 inch on each edge, and when these planks are placed edge to edge it will show a seam of about ⅛ inch in width. Draw a line through the center of the beams, and after cutting the end of one of the pieces of plank to fit the transom, bring the edge of the plank to the line and fasten. Add planks on either side until you have the entire deck covered, taking due care to draw them close together. Calk the seams, and pay them with thick lead paint. The filling of the seams with white lead putty and planing the deck can better be finished when you have the boat nearer completion.

Now start at the opposite end and fit in at the stem an oak breast hook or knee, fasten to the clamps, sheerstrake and stem. The plans show seventeen beams for the raised deck; these should be of oak, sided 1¼ inches, moulded 1¾ inches. They can be either cut with a crown of 5 inches in 7 feet or steam bent to that shape. The lower corners can be chamfered or rounded, and the beams finished smooth. The next thing is to make a strong back to set these beams on, and for this you can take a piece of board about ⅞ inch thick and about 8 to 12 inches wide; set this so that the upper edge is the height of the under side of the beams, fasten both ends securely and proceed to cut the ends of the beams into the clamp, fasten to same and to the heads of frames. Cut out the covering boards, which are to be ⅞ inch thick and 3 inches wide, spring around the edge of the sheer and fasten to the beams and sheerstrake. Mark a center line on the beams and proceed to lay the plank, which is to be of cypress ⅞ inch thick, 3 inches wide. Fasten them from above into the beams and also toenail, as this will tend to draw the planks tight and make them close on the under side. Plane the upper side smooth and then fit in the grub beam, which is to be of oak 2x6 inches, the upper edge of which is to be rabbeted to receive the staving at the after end of the house. This beam should be fastened very securely, as it will prevent the hull from spreading. The beams for the cockpit deck can be gotten out; these are to be of oak, sided 1½ inches, moulded 1¾ inches, spaced 18 inches or fastened to each alternate frame. It is a good plan to nail a strip of yellow pine or oak about 1x2 inches to the frames at such a height that the ends of the beams will rest on same, and the beams should be fastened to this strip as well as to the frames. Set stanchions under the centers of the beams to make them more rigid and to help support the cockpit deck. Lay the cockpit deck plank in strips of pine the same width and thickness as for the deck at the after end of the boat. If you intend to finish this deck in natural wood varnished, you will have to calk the seams and fill them with white lead putty or marine glue, if you prefer that material. Provided you intend covering the cockpit deck with canvas, the plank can be of either pine or cypress; tongued and grooved stock is better for this purpose than the square edged material if it is not to be calked. When you have finished laying the deck you can proceed to cut out the rabbeted pieces of oak. The curve of the staving is so slight that this strip can be worked out of straight stock and sprung to shape. Twenty feet in length will make enough for the two sides and across

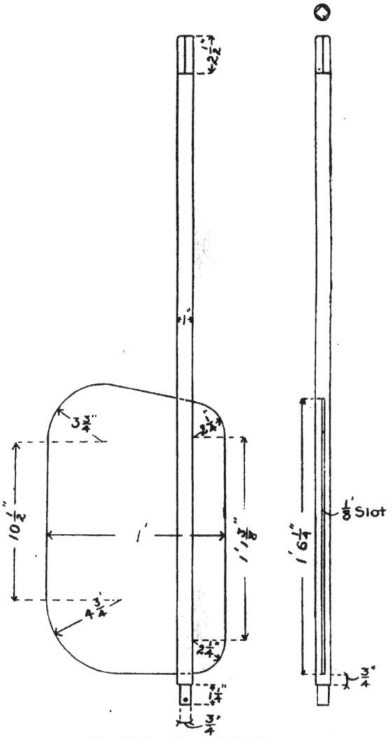

Detail Drawing of Rudder

the front of the seat. Take your spirit level and set it against the inside of the clamp, ascertain that it is plumb, then make a mark on the cockpit deck plank; continue doing this at intervals on both sides until you have a sufficient number of points to strike a line through with a batten and make a curve that corresponds exactly to the inside of the clamp. The line thus marked shows the outside of the staving, and as the pieces of oak you have rabbeted to receive the planking are 2 inches wide, you will have to set the outside of this piece of oak 1⅛ inches closer to the frames, as the staving is ⅞ inch thick, and you want to set the staving perpendicular. If the deck is covered with canvas, simply lay the oak pieces in thick white lead and fasten closely to make it watertight, but if you intend to have the deck bright, it would be advisable to set the oak pieces on a strip of calking cotton as well as using white lead. When you have set these oak pieces you can proceed to set up the cockpit staving, which is to be of cypress ⅞ inch thick, 2 inches wide, tongued and grooved, and the edges bevelled on the inside or face. Cut the lower ends so that they fit the oak pieces, set in white lead and fasten to same, and also fasten to the clamp. Leave the upper ends of these pieces of staving longer than you really require, as you can cut them off to a fair line when you have finished the staving. The staving at the after end of the cabin being of the same material can also be set in position. The lower ends are to set into a rabbet on the grub beam, and the upper ends are to be fitted to the under side of the raised deck plank, where it projects over the beam, and are to be fastened to the raised deck beam. A filler piece of spruce of some such material 1 inch thick and 2⅜ inches wide can be fitted in between the upper edge of the staving and the inside of the planking. The top of same is to be the same height as the under side of the cap, and when you have fastened in these pieces securely you can proceed to get out the cap. The specifications call for the cap to be of oak 1⅛ inches thick, 5 inches wide; this will allow for ½ an inch overlap on the outside of the planking and also the same amount on the staving. Round these edges and proceed to fasten in position. You will most probably find that you cannot bend this piece of oak to the proper shape without steaming it, and therefore it would be advisable to steam it in the first place. Put in plenty of fastenings, the filler piece and the edges of the planking affording good places to fasten to. The heads of the fastenings should be counter sunk and covered with wood plugs, as should all the heads of the large fastenings in any part that is to be finished bright.

The coaming for the cockpit hatch over the engine is to be of oak; work this out to shape as per plans, allowing for a waterway between the forward end and the grub beam, and fasten to the deck and beams. Rabbet a piece of oak 1½x2 inches to fit the upper edge of the coaming; this piece is to be used for the frame of the hatch. Halve the corners together and proceed to cover with ⅞x2-inch white pine strips as you did with the cockpit deck; if you desire to finish bright, calk and fill the seams and put a small oak ribbon around the outside to form a finish and cover the wood ends. If covered with canvas this ribbon should cover the fastenings.

The seat at the after end of the cockpit is not completed, and you can finish same before proceeding with the cabin. The specifications call for the top of this seat and the lazy back to be of mahogany, but it is simply a matter of choice what you use; mahogany stands the weather and warps but little, but I have seen some very fine looking seats and backboards made of cypress. Allow for part of the top of this seat to open, as there is valuable space under it that can be used for storage, and you might have to get under the after deck at some time or another to renew the tiller rope, etc. Set in the lead scuppers at the after corners of the cockpit, and it is practically completed. These scuppers should be about 1¼ inches inside diameter, the flange on the upper side should be set flush with the deck and the lower end of scuppers flush with planking. It will be necessary to cut a rabbet in plank to receive the flange. The fastenings in the lead scuppers should be of copper.

The towpost and forward bitts can be gotten out and fitted. These are to be of oak or locust 4x4 inches and 12 inches long. The part that fits the deck is to be trimmed to 3½x3½ inches; a hole is to be cut through same, and they are to be driven in position and secured on the under side with two oak or locust wedges. The towpost can be set up, but the forward bitt cannot be fastened in until you have covered the raised deck with canvas, which you can now proceed to do. The canvas should be 10-ounce material, 8 feet in width and 14 feet long. This must be well stretched and fastened closely around the edges with copper tacks. The usual method of laying a canvas deck of this description is to cover the upper side of the plank with a thick paint, stretch and fasten the canvas, and then dampen the canvas with a sponge and apply a coat of paint. The claim is that the moisture tightens the canvas and the coat of paint prevents it from relaxing. Owing to the diversity of opinions upon this subject, I am going to state that I either set the canvas in thick white lead paint and when stretched apply a good coat of oil paint, using no water, or else I cover the plank with Jeffrey's marine glue, and when the canvas has been stretched it is made to adhere to same by being ironed with hot flat irons. Men who are well versed in the handling of canvas or duck tell me that the oil (linseed) destroys the cotton, and they ought to know. The edges of the canvas, where fastened, should be covered with a 1¼-inch half-round moulding.

Now that you have the deck finished you can remove the piece of wood you used for the strong back and proceed with the interior.

Set the beams for the floor and lay the planking. There should be three strips through the center of the floor which will not be fastened to the beams but cut up in suitable lengths and cleated together on the under side. These you can remove at any time to clean out the bilge.

The blocks for the lights are to be of cypress, 1½ inches in thickness; the sizes vary from 10½ to 12 inches square, according to the diameter of the glasses. The diameter of glass in the clear is 8 inches for the two after lights on either side and 7 inches for the forward ones, if you intend using fixed lights. If you have decided to use composition open ports they should be smaller; purchase those with the round frames, as they do not have to be cut into the planking, and present a much better appearance than the hexagonal frame light. If you can obtain the use of a lathe or get the blocks turned, you will save considerable work, and while they are in the lathe you can cut a rabbet ⅜ inch deep on the face to receive the ceiling. Mark on the outside of the plank the position of the different lights and saw out the circle, then fit the blocks from the inside, cut to the shape of the planking and use plenty of screws or nails to secure them to the planking.

The specifications call for the ceiling to be of cypress ⅜x2 inches, tongued and grooved, edges bevelled. Start the first strip close to the under side of the raised deck clamp, fit around the rabbet in the blocks and fasten to same and to the frames. After you have fitted four strakes you will find that the next strake will have to be cut at the forward end for a certain distance, as it will bear against the upper side of the clamp, and the next two strips will be shorter in length for the same reason. If you fit them carefully you will find that it looks well. The proper way to ceil this space would be to take a spiling and shape each piece so that there would be the same number of pieces at one end as at the other, but so that they would decrease in width as they neared the stem. There would be little or no advantage in ceiling a boat of this description in such a manner unless you were particular, but that is the proper way to put in a ceiling. Work in the ceiling from the under side of the clamp to the upper edge of the bilge clamp and, with the exception of the toilet room, it will cover the inside of the frames. If you want to make the interior look well you can ceil from the under side of the bilge clamp in the toilet room to the floor.

The companionway slide, runs, door frame, etc., can now be gotten out. They do not need any explanation, as the plans show the section of the hatch, etc. The specifications call for them to be of mahogany, but if you prefer some other material, well and good—use it; but don't use any wood that is apt to warp out of shape when exposed to the elements or you will have a leaky companionway slide. The doors, to look well, should be panelled and made of 1⅛-inch stock with panels at least ½ of an inch thick.

The rubstreaks can be of oak or teak, 2 inches wide. If shaped as shown, it presents a better appearance than if it were half round. Fasten well through the planking into the frames and take care to have a fair curve, as the appearance of such a boat can easily be spoiled when the rubstreaks are set in position if they are not in a nice, fair line, and there is really no excuse for them being otherwise, as you have your sheer line to work to.

When you have arrived at this point you have practically completed the hull. Fit a piece of ¾ inch half-round brass to the stem, to protect it, and bore for the rudder port if you have not already done so. For the rudder port you will need a piece of 1 inch brass steam pipe 18 inches in length. Have a locknut thread cut in the lower end 5 inches in length, at the upper end you will require a thread 2 inches in length. For the stuffing-box you can purchase a standard 1-inch brass cap and also a small locknut of the same size. Bore the cap to the size of the rudder stock and it will make a good stuffing-box for such a rudder. I don't wish to convey the impression that a stuffing-box made in this manner is as good as having a proper one made, but it is something that can be purchased most anywhere, and therefore would appeal to the amateur. The port should be carefully threaded into the knee and horn timber, and if you cover the thread with white lead and have it fit tight into the wood it will not leak. Fig. 6 shows the dimensions of the rudder, and a simple way to make this is to buy a piece of hard rolled Tobin bronze ⅛ inch thick and cut it to the required shape. For the stock you will need a piece of Tobin bronze 1 inch diameter, 3 feet 8 inches in length. Have a machinist mill a slot in this stock to receive the blade; square the head above the stuffing-box to receive the quadrant and turn the lower end down to ¾ inch diameter for a distance of 1¼ inches to form a pintle for the shoe to hold the lower end of rudder stock. You can either have the machinist rivet the blade in place or do this work yourself, as it is not difficult. A rudder made in this manner is well suited for so small a craft. The shoe you can bend up out of a piece of brass or bronze, but it is better to make a pattern and have this

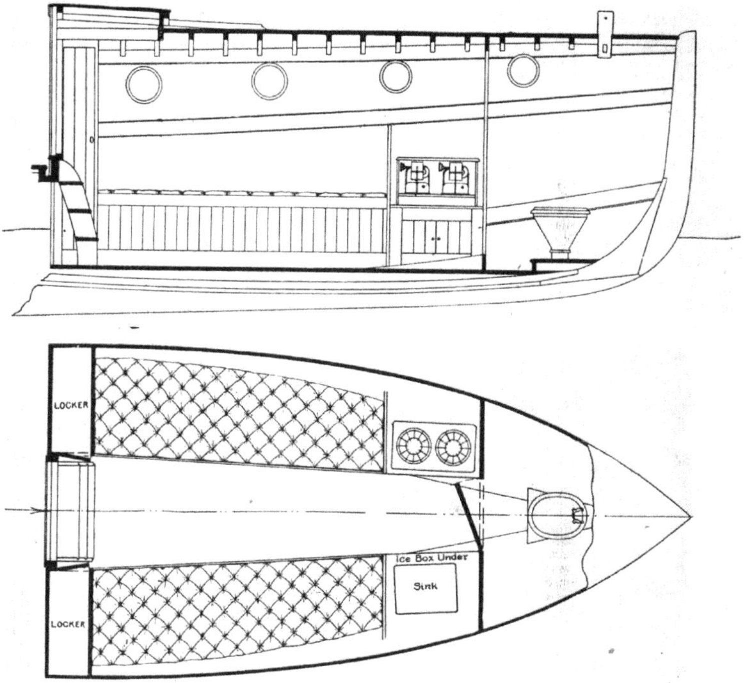

Another Cabin Arrangement for the Nock 25 Foot Cabin Cruiser

cast. The quadrant is a standard article with a radius of 12 inches and can be purchased from any well-known dealer in boat hardware. For the steering wheel you can use either a drum steerer or one of the auto steerers if you prefer to have a horizontal wheel; in either case you will need a tiller rope about ¼ inch diameter, and this can be led either between the staving in the cockpit and the frames or else under the beams of the cockpit deck. Wherever this wire rope, if you use wire, makes a sharp turn, it should run over a wheel with a diameter of not less than 2½ inches.

The interior can now be finished, and as I have shown two cabin plans, you can please yourself which you use, or perhaps you may have an idea of a cabin arrangement which would meet your requirements much better. Under any circumstances it is not worth while my explaining how to set up staving, bulkheads, seats, etc., for if you have finished your boat up to this point you will find that such work as this is of no trouble whatever. If you require any great amount of panel work, it would be cheaper to have this made where they have machinery, but I have purposely made the interior plain, thereby making it easier to construct.

A word about the engine: if you intend to install it, take just as much care in making up the joints on the piping, etc., as you would in making the joints in the hull, for it is of the utmost importance, and no matter how long it takes, if done properly, you will feel well repaid for your trouble. Use a seamless, drawn bronze or copper pipe for the gasolene supply, as this material does not deteriorate like brass pipe or crystallize like block tin pipe. The proper methods of installing engines have been thoroughly explained in MOTOR BOAT before in detail, therefore it would be useless for me to go into the matter. The gasolene tanks figured for this boat are cylindrical in form, 10 inches diameter and 48 inches in length. They are to be situated under the cockpit deck, well out on either side of the engine. A small water tank can be fitted under the after end of the berths in the cabin if desired, and a pump could be set up at the sink to supply the water for cooking and washing, and, if one required it, a small folding lavatory could be fitted in the toilet room.

The specifications mention the principal fittings, all of which can be purchased from any reliable dealer in boat hardware.

The finishing of the boat is quite important, the top sides and under body should be planed smooth and fair, then sandpapered and primed. Before the second coat is applied the first coat should be well rubbed down with fine sandpaper, and the same treatment to be applied every time before you apply a coat of paint. The under body should be treated in a similar manner, only in this case you should apply either copper or some anti-fouling paint. The *designed water line*, or the line shown on the plans as the L. W. L. is supposed to represent the line of the water when the boat is afloat, and it is a good plan to mark a waterline on a boat of this size 1½ to 2 inches above this line and paint up to this point with the same paint you use for the bottom, as it keeps the sides clean and adds to the appearance of the boat. This line should be marked on while the boat is in the stocks and scribed in with an awl, as it makes it easier to cut to such a line when painting.

All the bright work, such as rubstreaks, cockpit staving, companionway, etc., should be treated to one coat of filler, rubbed down and then have three coats of spar composition applied, each coat being rubbed down before the next is laid on. The canvas should receive one coat of canvas filler and two coats of deck paint. The interior bright work should have one coat of filler and two coats of a first-class interior varnish.

When you have completed the boat she can be removed from the stocks and launched, and there is no need for any explanation how to proceed to launch this craft.

Without doubt there are many minor things I have not mentioned that will have to be done, but I have gone into the matter in such a manner as I believe will make it plain to those who think of building a boat of this sort and have some fair knowledge of the use of tools.

Specifications for Construction of a 25-foot Cabin Cruiser

BY FREDERIC S. NOCK, EAST GREENWICH, R. I.

DIMENSIONS

Length, overall	25 feet	0	inches
Length, waterline	23 feet	4	inches
Breadth, extreme	7 feet	0	inches
Breadth, at waterline	6 feet	5	inches
Draught, to Rabbet	1 foot	4	inches
Draught, extreme	2 feet	1½	inches

MATERIAL AND WORKMANSHIP

In carrying out these specifications, there are to be used only the best materials and workmanship. Proper care to be given to the details of construction, fastenings, etc. All wood shall be sound, clear and free from all defects, all pieces to be cut fair with the grain, and all knees, etc., to be natural crooks, all fastenings not otherwise specified to be of galvanized iron.

KEEL

To be of native white oak, sided 3 inches, moulded as per plans and to be in one length, mill dressed to a uniform thickness and finished smooth.

STEM

To be of white oak, sided 3 inches, moulded as per plans, to be connected to the keel with an oak knee sided 3 inches, to be properly fitted and fastened with ⅜-inch diameter bolts clinched over washers, heads to be countersunk and covered with wood plugs.

SHAFT-LOG

To be of white oak, sided 3 inches, built up of two pieces, bored for shaft and to be securely fastened to the keel with 5-16-inch diameter bolts.

HORN TIMBER

To be of white oak, sided 3 inches, shaped as per plans, to be securely fastened through shaft-log and keel with 5-16-inch diameter bolts.

STERN POST

To be of white oak, sided 3 inches, shaped as per plans, to be carefully fitted and well fastened to the keel, etc.

TRANSOM

To be of oak, 1¼-inch thick, to be connected to the horn timber with an oak knee or hackmatack knee sided 2½ inches and fastened with ⅜-inch diameter bolts. Cheek pieces of oak 1 inch thick are to be securely fastened to the inside edge of the transom to form a back rabbet and fasten the planking to.

FRAMES

To be of white oak, 1⅛x1⅛ inches, spaced 9 inches center to center, frames wherever possible are to extend from gunwale to gunwale in one piece, and to be securely fastened to the keel, frames aft of forward end of shaft-log are to be mortised into a cheek piece of oak 1 inch thick which same is to be well fastened to the keel and shaft-log and the lower edge cut to form a back rabbet for the garboard.

FLOORS

To be of white oak, 1⅛x1⅛ inches, to be securely fastened to the keel and through the sides of the frames, floors aft of forward end of shaft-log to be sawn to shape, carefully fastened to the keel and through the frames.

BILGE CLAMPS

To be of yellow pine 1¼x4 inches amidship, tapered at

ends to 1½x3½ inches, length of taper 7 feet, to be securely fastened through frames.

Deck Clamps

To be of yellow pine 1¼x4 inches, upper and lower edges to be chamferred from forward end for a distance of 14 feet, to be securely fastened through frames and at each athwartship frame to have one fastening through clamp, frame and strake, heads of fastenings to be sunk in strake and covered with wood plugs.

Raised Deck Clamp

To be of yellow pine 1⅛x3½ inches, lower edge bevelled, to be well fastened to the frames and at each alternate frame to have one fastening through clamp, frame and sheerstrake.

Planking

To be of yellow pine in long lengths, to finish ¾-inch thick, where butts occur same to be reinforced with oak blocking and to be securely fastened, heads of fastenings to be covered with wood plugs. All fastenings to be of galvanized iron chisel point nails clinched on the inside of the fromes. Seams to be tight on the inside, and planed with a suitable bevel for calking, seams to be calked with yacht cotton, payed with white lead and finished flush with white lead putty, exterior to be carefully planed smooth and fair, and sandpapered before being painted, inside of plank to be coped to fit the frames wherever necessary.

Keelson

To be of oak or yellow pine, as desired, 2x4 inches, to be scarfed and fitted to knee at stem, and to be well fastened through floors and into keel.

Engine Bed

To be of oak, constructed to meet the requirements of the engine. Fore and aft bearers to be sided 3 inches, carefully fitted over frames and securely fastened to same, athwartship bearers to be sided 2½ inches, to be carefully fitted to the plank and securely fastened to the keel and fore and aft bearers.

Deck Beams

To be of oak, sided 1¼ inches, moulded 1¾ inches, spaced as per plans, to be cut with a crown of 1½ inches in five feet, ends to be mortised into deck clamp and well fastened.

Raised Deck Beams

To be of oak, sided 1¼ inches, moulded 1¾ inches, spaced 9 inches, crown of beams 5 inches in seven feet, and to be either sawn to shape or steam bent as desired. Ends of beams to be cut into clamp and securely fastened to same and heads of frames.

Raised Deck Covering Boards

To be of oak, ⅞-inch thick, 3 inches in width, sprung to shape and securely fastened to the beams, clamps and sheerstrake.

Raised Deck Plank

To be of cypress ⅞-inch thick, 3 inches wide, tongued and grooved and edges bevelled on the underside, to be well fastened to the beams, the top planed smooth and fair and covered with 10-ounce canvas properly stretched and bedded in white lead, edges to be fastened with copper tacks and covered with 1¼-inch half round oak moulding.

Deck Plank

To be of white pine ⅞x2 inches, to be well fastened to the beams, heads of fastenings covered with wood plugs, seams calked, payed with white lead and finished flush with white lead putty.

Cockpit Deck Beams

To be of oak, sided 1½ inches, moulded 1¾ inches, spaced 18 inches center to center, ends of beams to be fastened to the frames and supported on stanchions in center.

Cockpit Deck Plank

To be of white pine ⅞x2 inches, to be well fastened to the beams, heads of fastenings covered with wood plugs, seams calked with yacht cotton, payed with white lead and finished flush with white lead putty, or if desired, deck to be covered with 10-ounce canvas properly stretched and bedded in white lead.

Grub Beam

To be of oak, 2x6 inches, upper edge to be rabbeted to receive the ⅞-inch staving at after end of house, ends to be well fastened to the frames.

Cockpit Hatch and Coaming

Coaming to be of oak 1½ inches thick, cut with a rabbet on the upper edge for hatch as per plans, after end of coaming to be securely fastened to the beam, forward end to be cut with a waterway as per plans and to be securely fastened to the grub beam, hatch frame to be of oak cut with a rabbet to fit the coaming, and covered with white pine ⅞ x 2 inches, seams calked as per deck or covered with canvas as desired, if canvas covered there is to be an oak nosing around same to cover the fastenings.

Cockpit Staving

To be of oak or cypress, ⅞x2 inches, tongued and grooved, edges bevelled, to be well fastened to the clamp and lower ends set into a rabbeted piece of oak as per plans and securely fastened, forward end of cockpit seat and after end of house to be staved up in the same material, a filler piece of spruce or some suitable material is to be worked to shape and fitted between the upper ends of staving and strake and securely fastened, and to this is to be fastened the upper ends of the staving.

Cap

To be of oak, 1⅛ inches thick, 5 inches in width, edges rounded, to be fitted over top of staving and upper strake and to be well fastened, heads of fastenings to be covered with wood plugs.

Cockpit Seat and Lazy Back

To be of mahogany ⅞-inch thick, top of seat to be arranged to lift up, lazy back to be removable.

Cabin Doors, Frame, Companionway, Run and Slide

To be constructed as per plans of mahogany.

Forward Bitt and Towpost

To be of oak or locust 4x4 inches, to be 12 inches in length, 6 inches above the deck, and the part that passes through deck is to have a slot mortised through same and to be wedged up to the oak blocks with oak or locust wedges, there are to be two rivets in each bitt below the opening for wedges.

Rubstreaks

To be of oak, 2 inches wide, and about ⅞-inch thick, shaped as per plans, to be well fastened to the plank and frames.

Blocks for Portlights

To be of cypress, 1½ inches thick, shaped to fit the inside of the planking and to be securely fastened to same, face to be cut with a rabbet ⅜-inch deep to receive the ceiling.

Ceiling

To be of cypress, ⅜x2 inches, tongued and grooved, edges bevelled, to be well fastened to the frames and to extend from the underside of the raised deck clamp to the top of transoms, etc.

Cabin Floor Beams

To be of oak, sided 1¼ inches, moulded 1½ inches, spaced 18 inches, ends to be well fastened to frames and supported in center where necessary.

Cabin Floor Plank

To be of yellow pine ¾ of an inch thick, laid in strips 4 inches wide, the three center strips are to be cleated and cut up in short lengths to make suitable traps for access to the bilge, the balance of the plank to be well fastened to the beams.

Forward Bulkhead

To be of cypress staving, ⅞x3 inches, tongued and grooved, edges bevelled, door to be of the same material to be well cleated to prevent warping.

Center Bulkheads

To be of cypress staving, ⅞x3 inches, tongued and grooved, edges bevelled.

Transom Fronts

To be staved up with ⅞x3 inch cypress, mopboard 4 inches wide of cypress, and facing strip of either cypress or oak 3 inches wide.

Transom Tops

To be of cypress ⅞-inch thick, to be supported on beams, and suitable traps to be cut in same to admit of using space under berths for storage.

Galley, Etc.

Sides of lockers, front of stove locker, ice-box, etc., to be built up of ⅞x3 inch cypress staving, locker doors of same material, ice-box top, shelves, etc., to be of ¾-inch cypress, ice-box to be sheathed with ½-inch spruce and lined with zinc, stove space to be lined with zinc. An enameled iron sink 12x16 inches is to let into the top of the ice-box and fitted with a lead discharge pipe to drain overboard. Lockers on either side to be fitted with shelves, etc., to meet the requirements of the owner.

Companionway Steps

To be of oak or cypress, shaped as per plans and fitted with chocks for feet, also hooks and eyes for upper end, steps are to be removable to admit of starting engine.

Fittings, Etc.

To comprise:—One small yacht closet properly set and plumbed, one 12x16 inch enameled iron sink properly plumbed, one Khotal or Primus two burner galvanized iron frame, two round frame open port lights 6 inches diameter for toilet room, four 8-inch round frame fixed lights and two 7-inch ditto for galley and cabin. All interior hardware, such as catches, hinges, bolts, etc., to be of brass or bronze. Brass or bronze stem band, bronze rudder, bronze shoe, brass rudder port and stuffing-box, galvanized iron quadrant, galvanized steel tiller rope, galvanized iron wheel leads for tiller rope, brass and wood steering wheel, one pair of polished brass bow chocks, one pair polished brass quarter chocks, brass padlock and hasp, etc. Scuppers in cockpit deck to be of lead pipe 1¼-inch diameter.

Motor, Etc.

To be a double cylinder 2 stroke engine, 4 inch bore, 4½ inch stroke; weight complete not to exceed 350 pounds. All accessories for engine, such as piping, fittings, gasolene tanks of 20 gallons' capacity each, shaft, propeller, stuffing-box, wiring, coil, magneto, etc., to be furnished complete.

Painting, Finishing, Etc.

Entire interior of hull to receive one coat of priming paint before being ceiled, exterior of hull to be finished smooth, and carefully sandpapered and given one coat of priming paint and three coats of pure white lead paint to the water-line, underbody to receive two coats of copper or Anti-Fouling paint. All exterior bright work to be treated to one coat of filler and three coats of Spar Composition. House deck and all parts covered with canvas are to be treated to one coat of canvas filler and two coats of U. S. Deck paint. Interior bright work to be treated to one coat of filler and two coats of Cabinoleum or I.X.L. varnish.

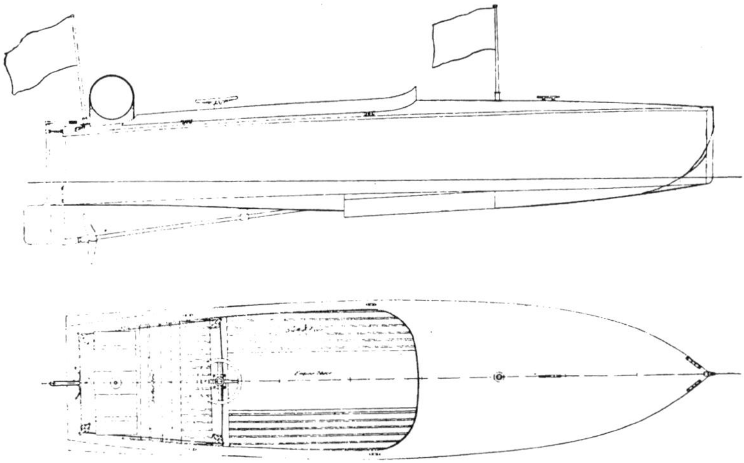

OUTBOARD PROFILE AND DECK PLAN

How to Build a Single Step Hydroplane

By George F. Crouch

THE racing results of the season of the past year have shown clearly that the hydroplane type of boat has come to stay. Every important race—referring, of course, to the speed boat classes—was won by some type of "hydro." Weather conditions had much less effect on their speed than we had been led to expect; and, taken as a whole, I believe that they were better performers in rough weather than the displacement boats of equal length and less speed.

Riding in a good "hydro" is a joy which can be found in no other sport. The little boats are so "alive," they respond so quickly, turn in almost their own length and flutter over the surface of the water in a delightful way. A displacement boat seems dead after one has become accustomed to the hydro. One misses the "pat-pat-pat" of the hydro as it glides over the ripples and one misses its stiffness. The displacement racer seems to heel down on her beam ends as if she were never coming back while the hydro keeps moving along on an even keel.

The progress of the last year in hydroplane design is a surprise to all who have closely followed the course of development. At first it was thought that the hydroplane must be marvelously light and every effort was made to cut down weight; hulls were pared to the limit, motors of the lightest possible types were chosen, and in some cases this weight reduction was carried to a point where light men were chosen for the crew. No doubt this was the correct thing to do with the hulls used, but the latest models show that fairly heavy construction and a medium weight racing motor can be used with almost equally good results as far as speed is concerned, and much better results when endurance and reliability are considered.

The little boat, which is the subject of this article, is of this latest medium weight type. The hull is not extremely light but the construction is such that it is unusually stiff and rigid. As may be noticed, the cross floors, the keel and the two fore and aft stringers form a strong truss even without the planking and the fore and aft edge stringers.

The hull is of the single step type, the step being formed by putting a metal plane on the hull after it has been completely planked. It is not necessary to explain the manifold advantages of an applied metal plane as any one interested in hydroplanes knows them. With regard to the hull construction it is not an easy matter to show a boat which will be easily built and yet possess all the strength, rigidity and lightness that the boat turned out by the professional builder will show. The general scheme which I have used is that of having no bent or steamed frames whatsoever. The moulds, instead of being made of rough material,

GENERAL DIMENSIONS OF HYDROPLANE

Length, overall	19	feet	4	inches
Beam, extreme (at planksheer, outside ½ round)	4	"	8¼	"
Beam, extreme (at chine)	4	"	4⅜	"
Depth of hull (forward)	2	"	2½	"
" " " (amidship)	2	"	9¾	"
" " " (aft)	2	"	3½	"
Draft of hull (at rest)	0	"	11	"
Draft (at rest, with 18-inch wheel)	2	"	5	"

are to be of selected spruce, nicely finished and put together as shown on the drawings of the molds or frames. These molds stay in the hull and take the place of the ordinary system of framing so care must be used in getting them out and in finishing them. In fact, throughout the whole construction anything put on the work stays there in the completed hull. There are no molds to rip out after the boat is planked, no ribbands to be used in "fairing up" and then taken off again.

Although the construction is fairly simple, it is a much more difficult boat to build than the *Water-Bug*, the plans and building instructions of which were published in Motor Boat for January 10 and 25, 1911, and I therefore recommend that boat for those who have never had any boat-building experience rather than for them to attempt this new one The troubles which the beginner would find in building to this new design are not so much in the hull itself—although the planking of concave "veed" bottom means work—as in the motor and drive installation I have counted on using a gear drive to the propeller shaft as this gives a good shaft angle and allows the

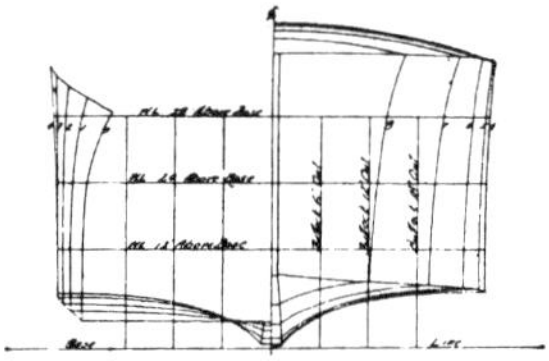

motor to be placed well aft. This motor position is of utmost importance in a boat designed to carry a fairly heavy motor. Since motors differ so widely in power, size and weight, I cannot give definite measurements for the driving gear and motor beds, so the builder must use his ingenuity and experience in following out in detail what I can only indicate in a general way

The motors which may be used in this hull range in power from 40-hp. to 120-hp. I would not advise using much less than 40-hp. and the weight of such a motor should be not over 600 pounds. The speed to be expected with such an equipment is 28 to 30 miles an hour, while with a 120-hp. motor weighing about 1,000 pounds the speed should be about 40 miles an hour. Any motor between these two should give proportionate results. Of course you would not use a 40-hp motor weighing 1,000 pounds for the weight should be in proportion to the power If a motor too heavy for the power is used, the boat will not "get up" but will plough along at canal boat speed The motors should be of fairly high speed capable of turning somewhere between 900 and 1,500 r p.m., and the driving gear can be proportioned to suit the motor chosen so that the propeller will turn 1,500 to 1,600 r.p.m.

I will not stop to tell you how to choose a place to build the boat nor what tools you will need, as I assume that you have some knowledge of boat building and know that small boats are always built under cover, that a hammer is used to drive nails, and so on.

The first step toward the actual construction of the boat is to get out the keel, stem, transom and the knees for connecting them together. The plan of the keel and the keel sections gives full information as to size and form of the keel. If possible, the keel should be in one length and the best material to use is white oak, although yellow pine might be used as a substitute in case it is impossible to secure oak Between molds No. 3 and No. 7 the keel has considerable bend in it, and it will be neces-

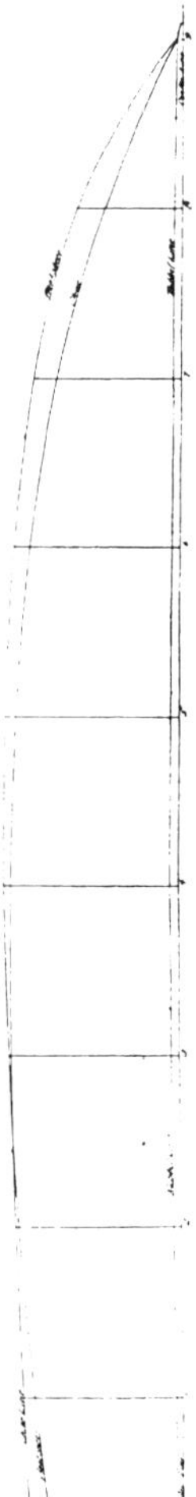

LINE DRAWING OF SINGLE-STEP HYDROPLANE
Lines are shown to outside of plank

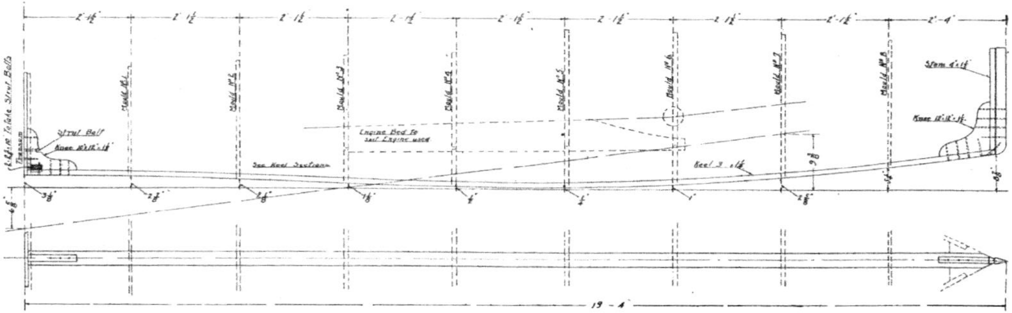

KEEL PLAN, SHOWING POSITION OF MOLDS

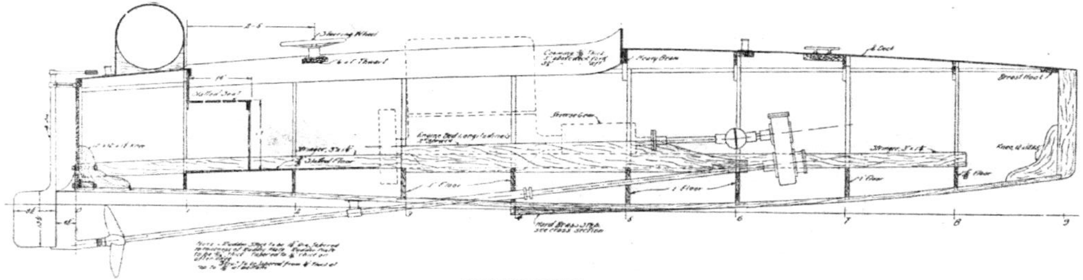

INBOARD PROFILE

sary to take the stiffness out of it by steaming in a long steam box or by pouring boiling water upon it until the required bend can be given to it. If you cannot secure a piece of oak long enough to make the keel in one piece, two pieces may be used and the joint between them stiffened by an oak butt strap about 2 feet long, thoroughly riveted to each part of the keel.

The stern is a piece of white oak shaped as shown on the detail plan. The rabbet should be roughly cut to form, as should that of the keel, leaving the finishing touches to be given after the frames have been set up and the boat is ready for planking. The knee joining the stem to the keel is sided 1½ inches, the same as the stem and may be made of hackmatack or even of straight grained oak, in which the oak should, of course, run diagonally between the stem and the keel. Note the stopwater where the keel joins the stern.

The transom should be made ready next. It is of ⅝-inch mahogany or oak shaped as shown in the transom detail. It is too deep to be made in one piece, so the separate parts must be joined together and then stiffened by vertical pieces of oak as shown. Do not cut the round at the top of the transom down to the line, but leave that until the deck is on.

In getting out the molds or frames, which is the next operation, you will find it a good plan to make full sized drawings of these molds on heavy paper or on a clean board before you start in. These full sized drawings will allow you to compare the shape of the pieces you are making with the required shape and to check the completed mold after you have riveted it together. This drawing should be made for both sides of the mold from the dimensions given on the mold or frame details. These dimensions on this drawing are given to the inside of the planking.

The molds themselves are of spruce, the pieces up the sides being 3 inches wide and ⅝ of an inch thick, the bottom cross floors are 1 inch thick in way of the motor and gear drive, and are ¾ of an inch at the ends of the boat. The pieces should be cut to shape and then riveted together as shown, using corner pieces of oak ½ inch thick. The molds which have deck beams may have these cut and fitted at the same time as the other parts of the molds. The molds on which beams are not required must be held across the top by a temporary cross piece in order to keep them from spreading or squeezing together. Do not cut any notches in the molds for the fore and aft ribbands, that covers up the seams between the planks until after the frame is set up, but the notches to take the keel, the chine and the clamp should be cut as shown on the drawings. When setting up the molds be careful to get them spaced just as shown on the plans, or else you will find that there will be trouble in getting the proper bevel on the molds after the plank edge battens have been run in.

After the various parts have been prepared as described, they should be "set up." I am not going into any great detail with regard to "setting up" the molds as the drawings show clearly the relation of the parts to each other. The boat may be built upside down or right side up equally well. Some builders will prefer the first method and others the second. Whichever you use, be sure that all parts are securely fastened and that the molds are all parallel and square across the center line. If you set the molds in an inverted position, which is probably the easier method for the amateur, the keel should be kneed to the transom as shown, and fit into the notches cut in the molds to receive it, and should then be bolted to the knee at the stern. Make sure that the bend in the keel does not force any of the molds out of position. The chine piece of 1 inch by 1 inch yellow pine or spruce should then be run fore and aft

OFFSET TABLE

	Heights Above Base Line							Half Breadths						
Stations	Bottom of Keel	Rabbet Line	Buttock 6" Out	Buttock 12" Out	Buttock 18" Out	Chine	Plank Sheer	Bottom of Keel and Rabbet	Chine	W.L. 12" Up	W.L. 20" Up	W.L. 28" Up	Plank Sheer	Stations
0	0-3-1	0-3-2	0-3-2	0-3-2	0-3-2	0-3-2	2-4-3	0-0-6	1-11-4	1-11-3	1-10-7	1-8-0	1-7-7	0
1	0-2-7	0-3-0	0-3-5	0-4-2	0-4-4	0-4-4	2-6-1	0-0-7	2-0-7	2-0-7	2-0-5	1-11-4	1-11-0	1
2	0-2-1	0-2-2	0-3-7	0-4-7	0-5-3	0-5-4	2-7-5	0-1-0	2-1-6	2-1-7	2-1-6	2-1-4	2-1-2	2
3	0-1-1	0-1-2	0-3-6	0-5-3	0-6-0	0-6-2	2-8-7	0-1-1	2-2-3	2-2-4	2-2-4	2-2-5	2-2-6	3
4	0-0-4	0-0-5	0-3-6	0-5-5	0-6-3	0-6-6	2-9-6	0-1-1	2-2-3	2-2-3	2-2-5	2-3-1	2-3-5	4
5	0-0-2	0-0-3	0-3-6	0-5-6	0-6-5	0-7-0	2-10-4	0-1-1	2-1-6	2-1-6	2-2-0	2-2-4	2-3-4	5
6	0-1-0	0-1-1	0-4-2	0-6-1	0-7-0	0-7-2	2-10-7	0-1-1	1-11-5	1-11-6	2-0-2	2-0-7	2-2-0	6
7	0-2-5	0-2-6	0-5-3	0-6-6	0-7-4	0-7-4	2-11-2	0-1-0	1-7-4	1-7-5	1-8-2	1-9-2	1-11-1	7
8	0-5-2	0-5-3	0-7-1	0-8-0	—	0-8-0	2-11-3	0-1-0	1-0-1	1-0-2	1-1-0	1-2-2	1-4-4	8
9	0-8-7	0-9-0	—	—	—	0-9-0	2-11-4	0-0-6	0-0-6	0-0-6	0-0-6	0-0-6	0-0-6	9

NOTE:- Frames spaced 2'-1½". For'd edge of stem 2½" forward of Station #9

Dimensions are given to outside of plank, in feet, inches, and eighths of inches

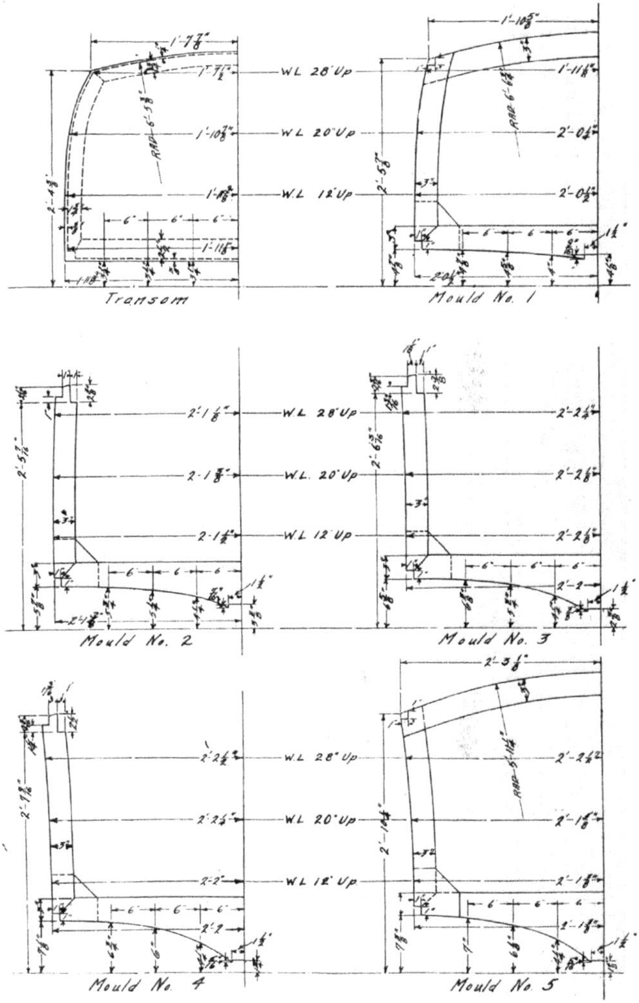

DETAIL OF MOLDS

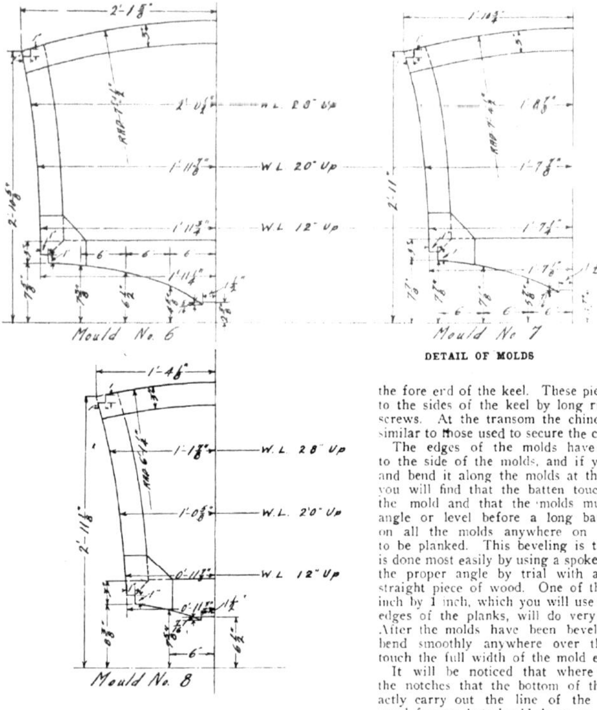

DETAIL OF MOLDS

in the notches at the corners of the molds and should be connected to the transom by knees and to each side of the keel at the fore end by oak blocks as shown in the sketch. The clamp is next run in by the same method, the size and material being the same as for the chine.

If the notches for the clamp, the chine and the keel have been cut square to the edges of the molds you will notice that the clamp, chine and keel will not fit evenly into these notches but will bear on one edge of the molds as shown in the sketch; particularly on the molds in the fore part of the boat. The notches must be trimmed down to the line shown in the sketch so that the clamp, chine and keel bear evenly on the molds and are then fastened in place by long brass screws or by long copper rivets if you so desire. The screws for fastening the chine and clamp should be about 2 inches long and those for the keel about 4 inches. The clamps on each side of the boat should be connected to each other and to the stern by a "breast hook," and should be kneed to the transom. The fore end of the chine is connected to the keel at the fore end by a triangular piece of oak which is shown in detail on the drawing of the fore end of the keel. These pieces should be fastened to the sides of the keel by long rivets or by long brass screws. At the transom the chine is fastened by knees similar to those used to secure the clamp

The edges of the molds have been cut off square to the side of the molds, and if you take a long batten and bend it along the molds at the sides or the bottom, you will find that the batten touches only one edge of the mold and that the molds must be trimmed at an angle or level before a long batten will bear evenly on all the molds anywhere on the surface which is to be planked. This beveling is the next operation and is done most easily by using a spoke shave cutting down to the proper angle by trial with a batten or long thin straight piece of wood. One of the pieces of spruce ¾ inch by 1 inch, which you will use later to run under the edges of the planks, will do very nicely for this work. After the molds have been beveled this batten should bend smoothly anywhere over the molds and should touch the full width of the mold edge.

It will be noticed that where the chine is run in the notches that the bottom of the chine does not exactly carry out the line of the mold on the two forward frames, but should do so on the other molds. The bottom of the chine should be planed to carry out the form given by the molds, and, of course, the outside of the chine should be flush with the sides of the molds.

The material for the battens, which are run fore and aft to cover the seams of the planking should next be prepared. For this you will need 16 pieces, each 20 feet long, of clear, straight-grained spruce 1 inch wide and ¾ inch thick. As may be seen on the midship section, the bottom planking consists of five planks on each side of the keel and the side planking of five planks above the chine. On the bottom, therefore, four of these battens are required on each side of the keel and these should be spaced so that they divide the bottom at frame No. 4 into 5 equal parts. I believe you will find it a simpler job to run these bottom battens about parallel to the center line of the keel, as shown in the sketch. The batterns should be held in the desired position by wire nails through them into the molds, and after marks have been made where the edges of the battens cross the molds they should be taken off and notches cut to receive the battens. Care should be taken to have the inside edges of the battens bear snugly against the bot-

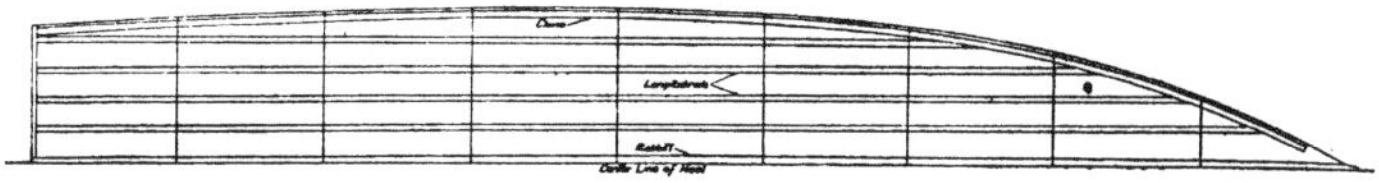

SKETCH SHOWING HOW BATTENS TO TAKE EDGES OF PLANK ARE RUN ALONG BOTTOM OF BOAT

tom of the notches, and to have the battens fair and smooth for the whole length.

Where the bottom battens meet the chine they should be fastened to it by a couple of screws or rivets, as shown in the detail sketch. At the stern these battens are let into the frame which is fastened to the fore side of the transom and fastened to it by a screw through the end of the batten.

The battens for the sides should be run in and notched into the molds in exactly the same fashion; they should be spaced to divide the side into five planks of about equal width, and should be fair and smooth throughout their length and should be flush with the molds, connect the pairs on the opposite sides of the boat where they come together at the stern by a breast hook and let them into the frame on the transom just as you did the bottom battens.

These battens, both along the bottom and the sides, should be fastened to the molds by 2-inch brass screws with the heads well countersunk into the battens.

When this is finished you will have a good idea of the form of the boat and will doubtless find it different from anything you have seen before. She is flat forward and flat aft with a fairly sharp section right amidships. The flat sections at the ends are so designed to give her a big lifting power. When running at full speed she really breaks the water with the sharp V-ed section of her forward plane, and this sharp section will make her easier and faster in rough water than she would be with a flat plane.

After trimming up the rabbet in the keel and stern, which may be done by laying a ⅜-inch thick strip along the battens and fitting it into the rabbet on the keel and stern, the boat is ready for planking. The material of the planking as called for on the midship section, is ⅜ inch mahogany, but clear white cedar or white pine could be used equally well. Mahogany is more expensive but it is also much stronger and will take a better finish than the other woods. It is also somewhat heavier but the difference in weight is much more apparent when the wood is dry than it is after the boat has been in the water for some time.

Start planking the bottom first and fit the garboards into the rabbet in the keel and have the other edge come on the center of the first batten. In order to do this you will have to take a spiling, using a thin board clamped or lightly nailed in place and marking off the points on it after the fashion which has been explained again and again in MOTOR BOAT. Owing to the curvature of the hollow V-ed sections you will doubtless find it necessary to make the garboard and the next plank somewhat thicker and then work this hollow into the planks with a plane. The planks should be fastened to the keels and to the battens with copper nails riveted over burrs and fastened to the mold and stern by screws. Set the heads of the rivets and screws flush with the outside of the plank as the plank is too light to allow any countersinking and plugging without weakening the fastening unduly.

The fore end of the garboard should be carried on out a little beyond the outside of the chine, and at the after end the garboard should butt against the flat of the transom. The same is true of all the other bottom planks. After the garboard has been fitted, and fastened in place, the next plank should be cut to the proper shape, and so on, each plank being cut from a spiling taken similarly to

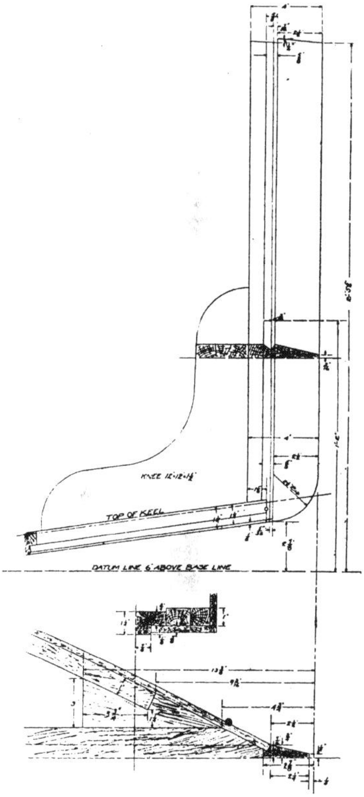

DETAILS OF KEEL, STEM, CHINE AND KNEES, SHOWING THEIR CONNECTION AT BOW

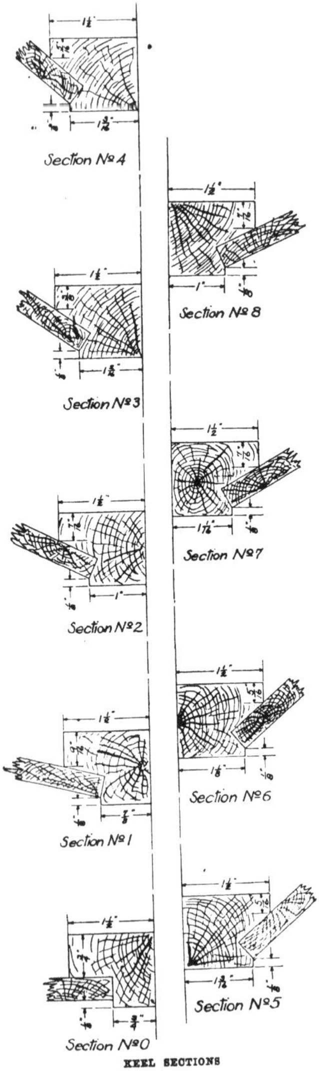

KEEL SECTIONS

the garboard and they should all run a very little past the outside edge of the chine at the fore end, and should butt up against the flat of the transom at the after end. Do not attempt to make the joints between the two planks too close, as the wood you are using is dry and will swell when the boat is placed in the water. Unless there is a little room left at the joint, the planks will be forced against each other, and will raise up in ridges. After each plank has been fitted, and before fastening it in place, it would be wise to paint thickly that part of the plank that bears up against the edge batten with white lead or a very thin coat of marine glue, as this will greatly assist in obtaining water tightness. The rivets in the battens along the plank edges should be spaced about two inches apart, the heads being flush, as before stated, with the outside of the planking.

After the bottom has been completely planked, trim the outside edge of the plank until it is flush and fair with the outside line of the chine, as the side planks are to be extended down to cover the joint between the bottom plank and the chine, and must, therefore, have a fair surface to fit against.

The side planking is put on, cutting the planks to shape from spilings, just as you have done on the bottom planks, allowing the lowest bottom plank to come down and cover the joint between chine and bottom plank, as stated above. The fastenings of the side plank are the same as the bottom plank, being screwed to the molds and stem, and riveted to the chine, edge battens and clamp.

After the hull has been planked, smooth up the entire surface, first going over all rivet heads with a file to get them flush with the wood, using emery cloth, to bring everything, rivets and wood included, to a smooth surface. This smoothing up is of the utmost importance, as much of the resistance of a hydroplane is due to surface friction. After smoothing up, give the hull a good coat of wood filler, allow it to dry hard, and then sandpaper again. Then give the bottom a thorough coat of varnish or bottom paint, whichever you desire. It would be wise to use at least two, or better three, coats of varnish or of bottom paint, but since this preliminary coat is given merely to allow us to put on the brass plane, and since the parts of the hull that are not covered by the plane can be finished and painted later, it would be a tedious and perhaps unnecessary delay to wait for the drying of these successive coats.

You are now ready to put on the brass plane. This is made of hard sheet brass, about one-sixteenth of an inch thick. The total length of the plane is 4 feet 5 inches. The position of the plane is shown on the inboard profile, extending from the fore side of frame No. 6 to the after side of frame No. 4. The appearance of the step formed by the plane is shown in a sketch. As may be seen, oak wedges, 2 inches deep at the step, are spaced about 5½ inches apart. The bottoms of these wedges are straight lines, and taper out to nothing at the fore side of frame No. 6. The wedge at the keel is made the full width of the bottom of the keel, for it will not be possible to get one sheet of brass of the width required. The two pieces of which the plane must be composed come together down the middle line of this middle wedge, and must be very thoroughly secured to this wedge with brass screws, spaced not over an inch apart. The wedges are fastened to the floors of frames Nos. 4 and 5 by long brass screws, put in from the outside, and between the frame screws are put down from the inside of the plank into the wedges. After the wedges are all on fair up the surface defined by the wedges by bending a thin strip of wood over them, and noting whether it lies smoothly over all the strips. Plane up the bottoms of the wedges until you can do this. The brass plates are next secured in place. At the fore edge they are screwed through the plank into the floor on frame No. 6, using 1½-inch brass screws spaced as shown in the sketch.

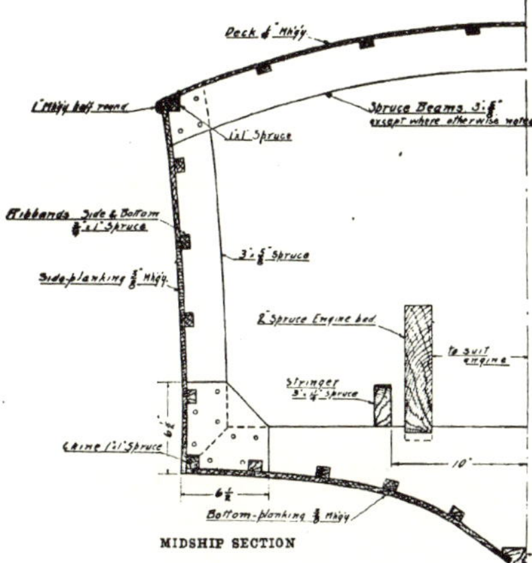

MIDSHIP SECTION

You are now ready to turn the hull over and put on the deck. The beams are shown on the molds, which give the curvature as part of a circle having a certain radius. These circular arcs can be easily drawn in by taking a long strip of wood, driving a wire nail into one end of it, measuring off the required radius from this nail, and drilling a small hole through which you may put the point of a pencil, then use this just as you would a compass to draw the arc. The top of the clamp and the top of the upper side plank should be cut to carry out the curve shown by the beams. The general scheme of deck construction is exactly that of the planking, but the material is somewhat lighter, and is of the size shown on the midship section.

The length of cockpit shown will do very nicely for a 40 or 50-h.p. motor, but, of course, if you intend to install a large motor, you will have to carry the cockpit one frame farther forward. This is something that you can readily determine for yourself, and depends on the motor chosen. Put in the spruce stringer to which the cockpit coaming is to be fastened, making it 1 inch by 2¼ inches, and running it in the notches inside of the upright parts of the frames. Run this stringer only between the beams at fore and after ends of the cockpit. Screw this stringer to each mold, and between each mold and the consecutive one put in two short blocks, filling the space between the clamp and this stringer. Then using a long one-eighth-inch rivet, make a thorough fastening at these points between the stringer and the clamp. At the fore end of the cockpit, pine cornerpieces about 2 inches thick should be cut out, to take the curve of the cockpit. If you are experienced, you could of course change the fore end of the cockpit to the more fashionable flared V type of coaming. Fair up the upper edge of the cockpit stringer to the lines indicated by the beam in the clamp, and then arrange edge battens of spruce 1 inch wide and half an inch thick, to go under the edges of the deck plank, as shown on amidship section. This deck is shown in fairly wide pieces, only one-fourth of an inch thick, and should be easily applied if the battens are spaced about as shown on the amidship sections, and are run fore and aft, practically parallel to the center line of the boat. Notches for

Do not let the fore edge of the plane into the plank, as this would cause weakness at this point. Simply file it down at a bevel; it will make very little resistance, as at extreme high speeds the boat will practically run on the plane itself and not on the forward part of the planking. The planes must be fastened to each of the wedges by brass screws, spaced about three inches apart, along the line of the wedge. All screwheads should, of course, be flush. Particular care must be taken to have the fastening at the forward edge of the plane very carefully done, for if it were possible for the water to force its way under this plane, it would soon rip it off. The after part of the plane at the step is left entirely open. When at rest, water may flow into the space between the outside of the planking and the upper side of the plane, but as soon as the boat is under way, it will drain out. If you wish to make a very fancy job, set the screw heads a little bit below the level of the brass, and then put a drop of solder over the heads. Next file this solder down flush, but there is no great need of doing this, if the slots in the screw heads are all placed in a fore and aft line, and the solder would be a great hindrance should you wish to remove the plane to get at the hull underneath. The plane should be finished by being polished by emery, either a fine emery cloth or emery flour.

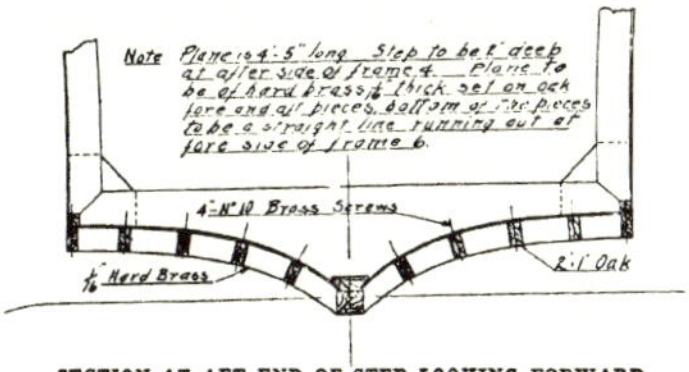

SECTION AT AFT END OF STEP LOOKING FORWARD

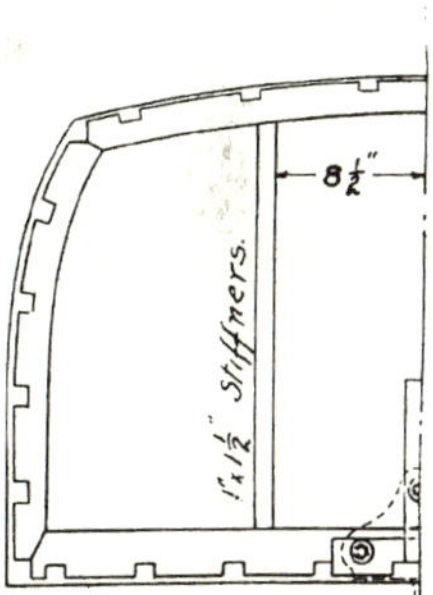

HALF SECTION, FORWARD
FACE OF TRANSOM

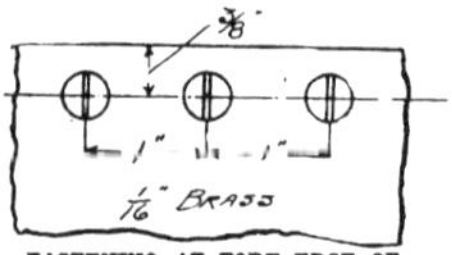

FASTENING AT FORE EDGE OF PLANE

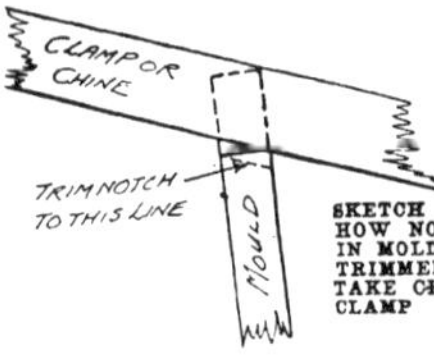

SKETCH SHOWING HOW NOTCHES IN MOLDS ARE TRIMMED TO TAKE CHINE AND CLAMP

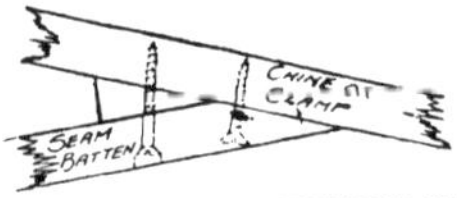

SKETCH SHOWING CONNECTION OF SEAM BATTEN TO CHINE OR CLAMP

the battens should be cut after the method used for the edge battens of the plank. The deck plank should be fastened with screws and rivets similar to the method used on the planking. Carry the planking to the outside of the side plank. Then trim it off smoothly and fairly with the side, covering the joint with a 1-inch half round of mahogany. A deck built after this fashion is light, strong and tight, but is not as handsome as the canvas deck, or the deck laid in narrow pieces.

The coaming is of oak or mahogany, 3/8 of an inch thick. If pine or cedar is used for plank and deck, oak might be used for the coaming, but it would, of course, be impossible to use oak on a mahogany planked boat. The coaming should be 4½ inches above the deck at the fore end, 3½ inches at the aft end. The detail of fitting the coaming is extremely simple, and space prohibits going into it in this article.

Slatted floors, about 3/8 of an inch thick, are shown on the plan. It is advisable to make these floors so that they may be removed in sections, as in this way you will be able to get at the bottom of the boat without ripping up the floor. The steering wheel of the simple drum type is shown placed on a thwart, an arrangement which gave unusually good satisfaction in *Water Bug* and is reproduced here. If the motor is exceptionally long, it may perhaps be necessary to dispense with this thwart and use a vertical post with a horizontal steering wheel

In getting in the shaft line, the approximate position of the shaft hole is readily seen from the drawing. Bore a comparatively small hole at about the position shown, putting on a small block on the keel to start the hole through the shaft at the bevel. Then from a point at the proper distance below the transom, as indicated on the drawing, pass a wire through this hole to the point shown on the mold as indicated, stretching the wire very tight. Then cut out around this hole, until it is large enough to clear the propeller shaft that you intend to use. The shaft line shown on the plans will take an 18-inch or 19-inch propeller. If you intend to use only a 16-inch propeller, it is advisable to decrease the shaft line from the bottom of the boat at the transom by about an inch. Use your common sense in making these allowances, as it is impossible to design a hydroplane to carry any old size of motor, and have everything work out to certain fixed dimensions. The motor bed is formed by fore and aft spruce stringers, set on the floor inside of two 3-inch by 1¼-inch spruce stringers; the lighter stringers run the full length of the boat, and are fastened to the floors by long rivets running clear through the stringers and through the floors. The heavier bed stringers are notched over the floors and are also riveted to them, but extend only the length required by the motor and reverse gear.

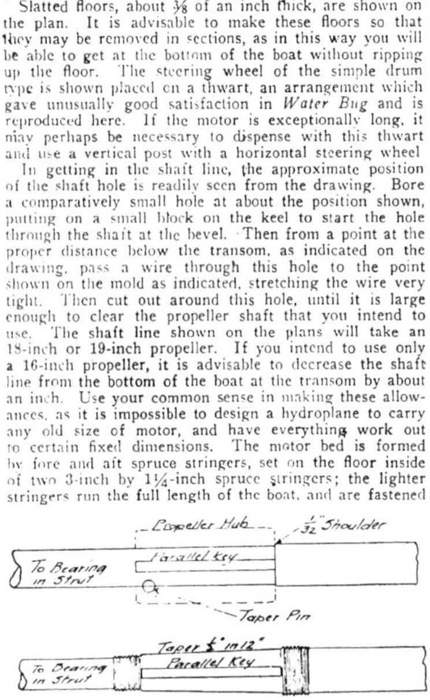

METHOD OF SECURING PROPELLER TO SHAFT

The transmission gear to drive the propeller shaft is a thing concerning which it is almost impossible to give definite detailed dimensions, unless the motor is known, its power known, in order to estimate the stresses which the gear must withstand. The gear should be made by an experienced machinist of the very highest class, one accustomed to high grade automobile construction will know about what is required, as this transmission is similar in general principle and in the degree of skill required in its manufacture to that used on cars. The case should be of aluminum, about one-fourth of an inch thick for a 50-hp. motor and about 5/16 of an inch thick for a 100-hp. motor. Lugs should be cast at the sides to carry the gear on a foundation built to receive it. This foundation, although it need not be extremely long, should be very substantial, as all the thrust of the propeller comes upon it, and besides that it is under considerable twist, due to the propeller and motor torque. The casing should have both the front and back all in one casting, as a much more accurate machine job can be obtained, than when one of the faces is made as a cover plate and bolted up against the rest of the casting If made as suggested, a cover plate at the top serves to introduce the gears, through which the shafts are slipped afterward. The drawing shows clearly the type of transmission gear advised. It is, of course, of the ball-bearing type, using annular ball-bearings of liberal size. The propeller shaft is fitted with a ball-thrust bearing for both reverse and ahead motion. In the gear shown, the upper gear on the motor shaft has 37 teeth of six pitch; the lower gear has 36 teeth of 6 pitch. This gear is suitable for about a 60 hp. motor. The gears have 1½-inch face, and the shafts are about 1 3/16 inches in diameter. The gears are held on the shafts by parallel feather keys. The casing is of aluminum. ¼ of an inch thick. It should be noted that a filling plug for oil is placed at the top of the casing, a drain plug at the bottom, and an oil level plug at the side. For the larger powers it would be advisable to use gears of 5 pitch instead of 6 pitch, and always make the number of teeth on the two gears so that they are numbers that are prime to each other. If you do this, then every tooth on one gear will eventually come into every space, in turn, on the other gear, so that wear will be evenly distributed. The gears wear in and run more quietly after having been in use for a time, than they do when new. If the gears have a number of teeth, such that they are divisible by the same factor, for instance, 48 teeth on one and 32 teeth on the other, both being divisible by 16, this distribution of wear is not accomplished; the gears wear unevenly, and become noisy. The gear material, the shafts and in fact every part of this transmission must be of the very highest quality. Special alloy steel should be used for the gear blanks, and after being accurately cut, they should be hardened. In designing the case for the gears, be sure to leave ample clearance all around the gears, in order to prevent your transmission from acting as a gear pump.

A universal joint is shown between the motor and the gear. This should be of ample size and is quite safe in this position, as it takes no thrust and is run at a comparatively small angle. Of course it could be eliminated, provided bevel gears at the proper angle were used in the transmission instead of spur gears, but bevel gears are much more expensive, are much more difficult to fit properly, and are apt to run noisily. Then, in addition, the universal joint provides for a certain amount of flexibility between the motor and transmission, which is very useful in hydroplane construction.

The strut and rudder present no unusual features, except that the rudder is hung at the bottom of the strut and at the top of the transom. The sizes are indicated at the joints. The strut is bolted through the back of the transom, one bolt at the back of the large palm going through the knee between the transom and keel. The bottom of the palm is wide, and has two bolts, the centers of which are about six inches apart, and go through the transom into an oak chock about 10 inches long and 2 inches deep by 2½ inches in a fore and aft direction. These bolts should be about ⅝ of an inch in diameter, and should be of bronze. The nuts should be at the outside, so that the strut may be removed, without crawling in under the deck. Sketches show how the propeller wheel may be applied to the shaft, either using a straight bore through the propeller, with a small shoulder at the fore end of the hub, and securing the wheel by a key and taper pin, or by using a tapered bore and nuts at the fore and aft ends of the propeller hub. The straight shaft and taper pin is much simpler, but it is much more difficult to remove a propeller than when the tapered bore and nuts are used. The water intake should be carried near the garboard, a little forward of the step, going down through the brass plane. An alternative arrangement would be to have the water intake project just a little below the plane coming down through the hull, just aft of the step.

For those who desire to work out their own construction, using perhaps a lighter, simpler scheme, the line drawing and offset table are given. If a big, heavy motor is to be used, the arrangement will work out better if the motor is placed aft and the crew forward of the motor, an arrangement which possesses many advantages; but in working out such an arrangement, it is necessary to know the definite motor weight, and should any of the readers desire to use this arrangement, I shall be glad to advise them where the center of the motor weight should be placed if they will furnish the necessary data regarding the motor sizes and actual motor weight.

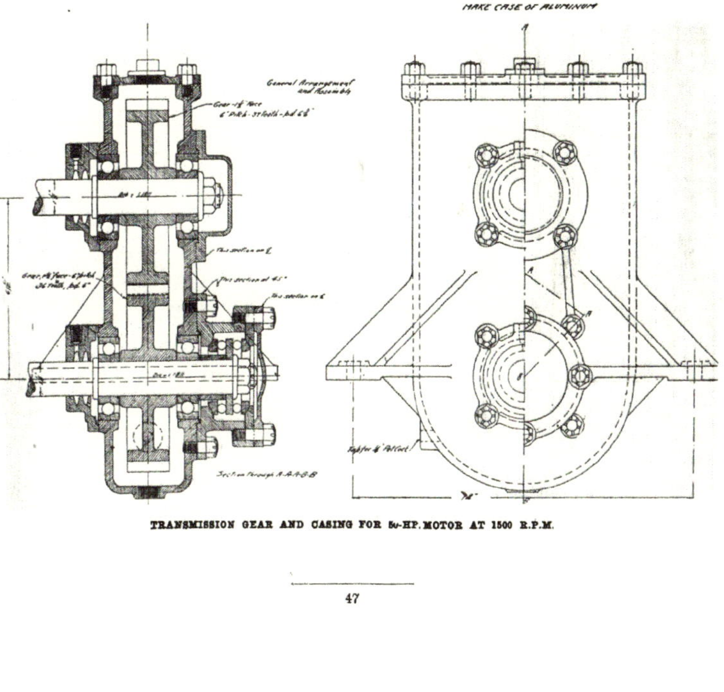

TRANSMISSION GEAR AND CASING FOR 50-HP. MOTOR AT 1500 R.P.M.

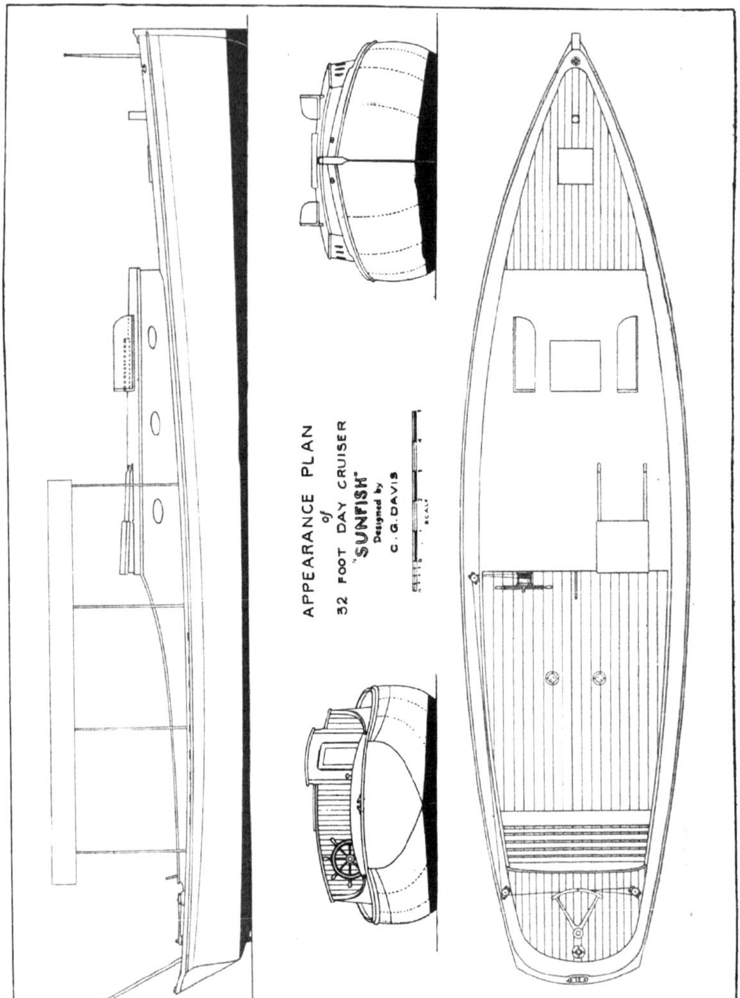
APPEARANCE PLAN
of
32 FOOT DAY CRUISER
"SUNFISH"
Designed by
C. G. DAVIS
SCALE

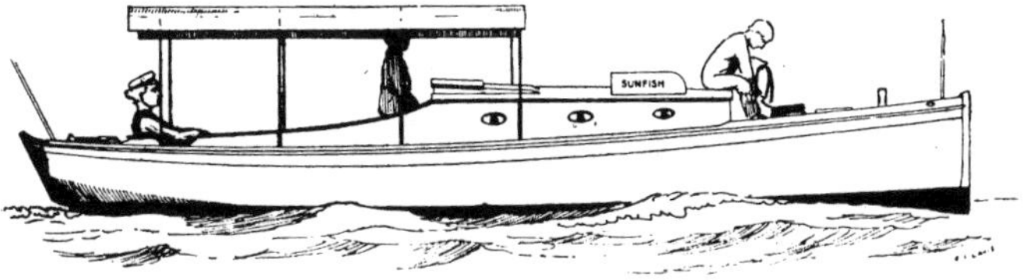

How To Build "Sunfish"

By C. G. Davis

WE have told how to build short, big-bodied motorboats where all style has been eliminated in order to give the biggest, roomiest boats for their size with no regard particularly for any set style, but now in *Sunfish* we are going to give instructions for building a good-looking day cruiser. By day cruiser we mean a boat that has deck room enough to enable one to move about, a big cockpit, a cabin top at a convenient height to sit upon and enjoy the breeze, and yet with cabin space enough for two or even three to sleep if they want to go cruising. But she is not like a raised deck cruiser—a house inside

With such a boat as *Sunfish* one can run up to a float and she is not so high-sided that you will need a side ladder to get ashore; you can easily jump or step from *Sunfish* onto an ordinary float. Such a boat does not chase all over the creek at her moorings, but will ride head-on at anchor, and though some may call her a bit old-fashioned, if she suits the purpose better than a raised decked boat why not have her so. Both bow and stern lines can be handled easily on such a boat.

Her model is just a clean, easy lined hull—nothing extraordinary, no startling features to attract attention, but a good sensible hull, plain and useful, yet of rather graceful proportions. On a length of 32 feet overall she is 29 feet 10 inches long on the waterline, 8 feet beam, and draws 2 feet 1 inch of water. She has about 2 feet freeboard at the lowest point, about the middle of her cockpit; 2 feet 7 inches at the center of her stern, and with the spray boards on her forward deck is 4 feet 3 inches above the water at the stem. She will float a total weight of 5,400 pounds at the waterline as drawn, her center of buoyancy there being just 6 inches aft of mold No. 5, or 16 feet 6 inches from the edge of the stem. She will swing a 20-inch propeller, and with a 10 to 15-hp. motor she will make a good 8 or 9 miles an hour. She will prove an easy boat to steer on a straight course, and yet will be prompt to answer her helm in turning.

Nobody builds a boat nowadays as they used to. Lumber can now be ordered at the lumber yard or saw mill in the sizes desired, and you don't have to hew and chop them out by hand, so the tools needed are mostly just a carpenter's outfit. I don't mean by that just a hammer and a saw, but such a kit as every carpenter is supposed to have. Such tools as the oldtime broad axe and whip saw are not required.

You can order a stick for the keel and get it already dressed—as they term planed lumber—to the size desired, but let me warn you right now, if you do order it dressed be sure to mark it down in big letters that you want the keel to be 3 inches by 4 inches *after* it is dressed. Otherwise you will get a stick that was 3 by 4 in the rough, and it will be 2⅞ by 3⅞ when you get it. The stick for this boat's keel must be 28 feet long and good for every inch, not a 28-foot piece with a foot of the end bad.

For the stem you want either an oak or a hackmatack knee, square or a trifle out square in its crook, 3 inches thick, without any skewgee or twist to it, with one arm 5 feet long, the other 3 feet, and thick enough in the throat to allow your stem being cut from it. It is safer to wait until you have drawn out the shape of your stem and made a ½-inch wooden pattern of it. By trying this pattern on the knee as you are selecting it, you can see if it is large enough or not.

There is one thing particularly needed in building any boat and that is a clear head. Stop and think out your work and don't believe the time spent in planning and laying out the work carefully is lost. It's all simple enough if you don't try to go too fast and get all confused. In laying out the stem, as an example, the outline of it is simple enough, but to tell how to bevel it off looks puzzling to the novice at first. Look at the plan showing the waterline's shape. As each waterline ends forward at the stem it comes in at a different angle. If you have laid the boat's lines down full size on the floor you can, with a bevel square, set that instrument or tool to that bevel and cut the stem until it fits. Each waterline from the deck down gets sharper and sharper. By spacing off these waterlines on your wood you can cut at each until you have it beveled to just what the lines call for. Don't bring the edge of the stem to a feather edge, but have it about ¾ of an inch wide to take a metal stem band.

The rabbet for the ends of the planking can be cut the same way by the use of the bevel or by taking a little piece of ½-inch pine board about a foot long and 3 inches wide and cutting a notch so half of one edge is ¾ inch wider than the other half. This ¾-inch projection represents the thickness of the planking. Chisel out the rabbet until this template fits on the face of the stern and the notched part fits snug in the rabbet. Another way is to wait until the molds are all set up and then bend a batten around them and cut the rabbet so the end of this batten fits true in the rabbet. The only objection to this is that it is more difficult to work in that position, standing upright, than it is where you can lay the stem flat on a floor or over a pair of wooden horses and sit on it and chisel out the rabbet.

The after deadwood can be made either in one piece or built up of smaller ones. If cut from one piece, which is more desirable, it takes a piece of 4-inch wood 18 inches wide and 6 feet 8 inches long. If built up of several pieces the upper part can be made from a 4 foot piece of 4-inch by 6-inch oak, and the shaft log from a 2-foot piece of 4-inch by 6-inch oak, and the deadwood below it from a 4-foot piece of 4-inch by 6-inch oak. The three must be jointed to a perfect seam where they meet and bolted to-

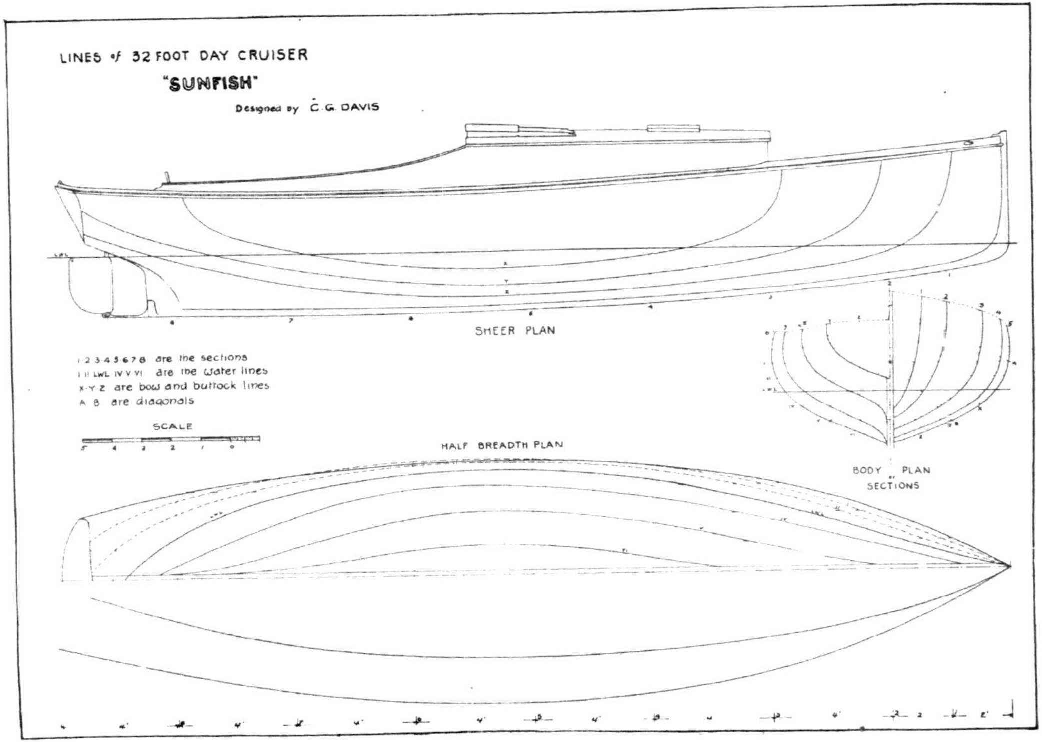
LINES of 32 FOOT DAY CRUISER
"SUNFISH"
Designed by C. G. DAVIS
SHEER PLAN
1 2 3 4 5 6 7 8 are the sections
I II LWL IV V VI are the water lines
X Y Z are bow and buttock lines
A B are diagonals
SCALE
HALF BREADTH PLAN
BODY PLAN or SECTIONS

gether with rods of ½-inch diameter galvanized iron or copper. You can buy this rod iron in 12 to 14-foot lengths and also the clinch rings that go over the ends where you rivet them up, but be sure to get wrought-iron clinch rings and not the brittle cast-iron ones.

In some localities it may be difficult to obtain a knee large enough to cut the stem from it. If so, it can be built up in two pieces just as the after deadwood, using a straight piece of oak 5 feet long, a foot wide and 3 inches thick for the stem proper and back of it a small knee about 2 feet long on its arms, as shown in the accompanying sketches.

It is to be supposed that a man who undertakes to build *Sunfish* has had some experience in the use of wood-working tools, and that he will know enough to be able to bore a bolt hole without choking his auger and in jointing up the deadwoods will square up the edges always from the face side so that when the various pieces come to be bolted together they will set true and level one on top of the other and not be canted or staggered out of the vertical. Such A, B, C principles a man is presumed to know when he tackles the building of this boat. The short sternpost is fitted dovetail to the after end of the shaft log so that the lag screws that are to hold the stern bearing will have crossgrained wood to hold to instead of end grain.

The bore of the shaft hole is so short that there should be no difficulty experienced in getting it through a solid log and so do away with the seam along the line of the shaft that would be there if the log were made up of two pieces with the shaft hole gouged half out of each. That is the way they are often built where there is a long deadwood to go through and in attempting to bore which the auger will often run off to one side or the other. Here the hole is only 22 inches long in the wood, a very easy job to bore.

As you work out each piece, scratch center marks and be sure that you set these marks all true when, after painting the two faces that come together, you rivet the stem and deadwood to the keel. Countersink the bolt-heads on the underside of the keel far enough to get a wooden plug dipped in white lead over them and so leave a flush, smooth job on the outside.

With the keel, stem and deadwood all together we have the backbone of the boat ready to set up and as the fairness of the boat depends on her being held rigidly to the desired shape while in the course of construction, be careful to get the shores, or short posts of wood that are to hold her keel, true to the measurements given above the floor—and don't trust to the floor's being true; stretch a chalk line very tight and measure up again to see that all is right before you set the keel up on them.

Set all the shores to a chalk line snapped down on the board floor if you are building her in a shed; if out in the weather, first where each shore is to come shovel away the loose top soil and sink a "deadman" just as a railroad tie is begged in the ground, tamp it down solid and then set your shores up on these. It is a great handicap to have to build out of doors. In a shed or shop you can run the braces to hold the head of the stem plumb

Table of Offsets
for the laying down of the lines of "SUNFISH"

Heights	Stem	1	2	3	4	5	6	7	8	Stern
Sheer	6-5¾	6-3¼	6-¾	5-8¾	5-5	5-2½	5-⅞	5-¾	5-1¾	5-3¾
Keel		2-1½	1-10	1-5	1-1¾	-11¾	-11	-11	-11	
Buttock X		5-2½	2-10	2-1	1-8½	1-6	1-6½	1-11½	2-9¾	3-9
" Y				2-8½	2-1¾	1-10⅝	1-11⅝	2-5½	3-4½	4-7¾
" Z				4-3¼	2-9½	2-5⅝	2-6½	3-1⅝		
Widths										
Deck		1-¾	1-10¾	3-1	3-8½	3-11½	3-10½	3-6¼	2-11¼	2-2
Water Line-I		-9½	1-7½	2-11¼	3-8¼	3-11¾	3-11¼	3-6½	2-9	
" II		-8¼	1-4¾	2-8¾	3-6¼	3-10¼	3-9⅛	3-4	2-3	
L.W.L		-6	1-1½	2-4¼	3-2½	3-6⅝	3-6¼	2-11	1-4¼	
" IV		-2½	-8½	1-9¼	2-7¾	3-¾	3-0	2-1½	-5	
" V				-10¾	1-8½	2-2¼	2-1	1-½	-1⅝	
" VI					-6	1-0	-11	-3¾		
Diagonal-A		1-1¼	2-1	3-5	4-1¾	4-5⅜	4-5	4-0	3-3½	
" B		-11	1-6¼	2-3	2-8¼	2-10¾	2-10¼	2-4¾	1-7	

Molds are all spaced 4 feet apart—except No. 1 which is 2 feet oft of stem and 2 feet ahead of Mould No. 2. Water lines are 6 inches apart, the L.W.L. being 3 feet above the base line from which all heights are measured. Buttock lines are spaced 12 inches apart. Diagonal "A" is 6 feet above base line—diagonal "B" is 4 feet up. A cuts water line No. I 4 feet out, B cuts water line V, 2 feet out.

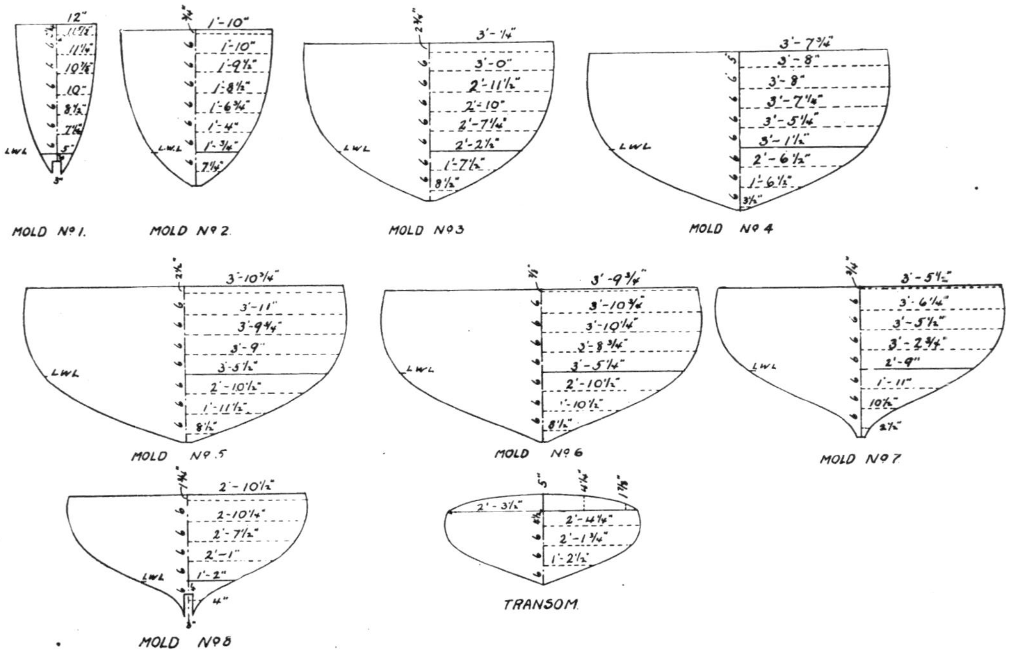

Measurements for laying out the shapes of the eight molds and transom used in the construction of a "SUNFISH"

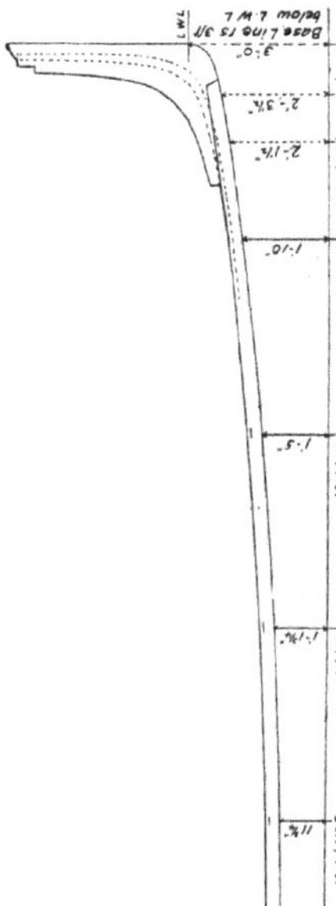

Plan showing the height for the blocking under the keel at each mold station - Molds are 4 feet apart

up out of the way to the rafters overhead, which you can't do on the ground. It prevents a lot of stumbling and dodging around the shores. Give the keel a coat of lead colored paint to preserve it where it is securely braced up plumb and true.

Then get out your molds and as these are only temporary a cheap grade of pine about an inch, or better yet, an inch and a quarter thick can be used. To shape all these by hand with a drawknife and saw is a tedious operation; if there is any place where you can get access to a band saw for an hour or so you could easily saw out the various pieces and put them together at home. Where you have to join two pieces together butt one against the other and then nail or screw with iron screws a cleat across the two to hold them. Keep all cleats and braces on the same side. Let the top end of each mold extend up four to six inches above the true sheer line so that you can run a batten up above this line and can leave it there until after you have the topstrake on to keep her deck edge fair and true. It is not necessary to bevel the molds; you do that by setting each of the molds forward of the center, 4, 3, 2 and 1, so the smooth side of the mold faces aft and is just flush with the mold marks on the keel, and the after ones 6, 7 and 8, the reverse way; by this the smooth edge represents the true shape required and when you put the battens on chisel away and bevel the mold's edge until it fits flat against the mold.

Be very careful to set each mold exactly to its mark and to set it and brace it perfectly plumb. If the under side of the cross-spall—the wooden brace across the top of the mold—be planed up true before it is fastened to the mold it will be found of great assistance in setting the mold level. You can hold a spirit level up under this edge and tap the mold to one side or another until the bubble sights true in the center of the level, and the plumb bob, hanging from the center marks on the cross-spall, is plumb over the center scratch line along the top of the keel.

Keep all the braces you can up overhead so that they will not interfere with your working around the boat, and cross them X shaped, as they go up to the rafters; by this arrangement you can get a more rigid brace than if they simply go straight up from the head of the mold. The whole business could swing like a parallel ruler that way, but the X brace is firm.

Before you can run the ribbands around you need the transom, the shape of the face of which is given along with the mold shapes. Do not cut it out to this shape however, as, due to the bevels, it needs to be an inch and a quarter wider on the bottom edge but no larger across the top where it bevels under from the line, and around the edge it may even take more. With sufficient wood left outside the line you can, after bending the curvature in the face of the transom, clamp-screw it to the small knee and by bending battens around the molds cut and fit it accurately. All these bevels could be laid out on the floor in the full-sized drawing, but it goes into a lot of projection and drawing, too difficult to attempt to explain here. MOTORBOAT HANDBOOKS, volumes 1 and 2, will help the amateur over a number of difficulties, as there are special articles in these books on laying down a set of lines, how to project a transom, an explanation of the meaning of a Table of Offsets, an article on how to cut the rabbet line, etc.

The transom is shown drawn to the outside of the planking. If you are going to let the plank fit flat against the edge of it you will have to take off the thickness of the plank from the shape shown, but as that does not look very neat on a job of this kind, due to the curve in the face of the transom, it would be better to bevel the front edge of the transom so that the plank ends make a seam right around the corner and are fastened to oak backing pieces screwed fast to the inside of the transom around the edges

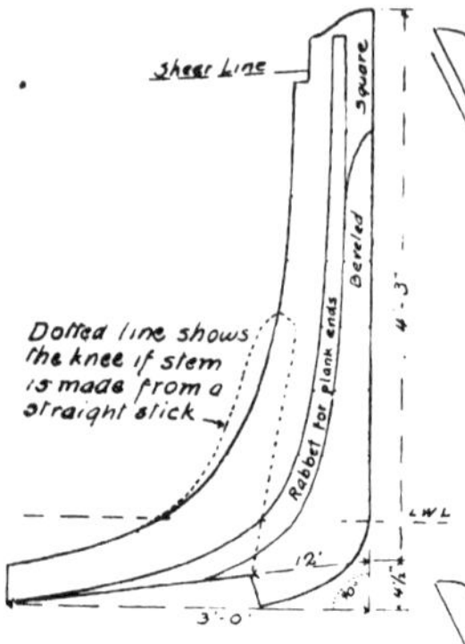

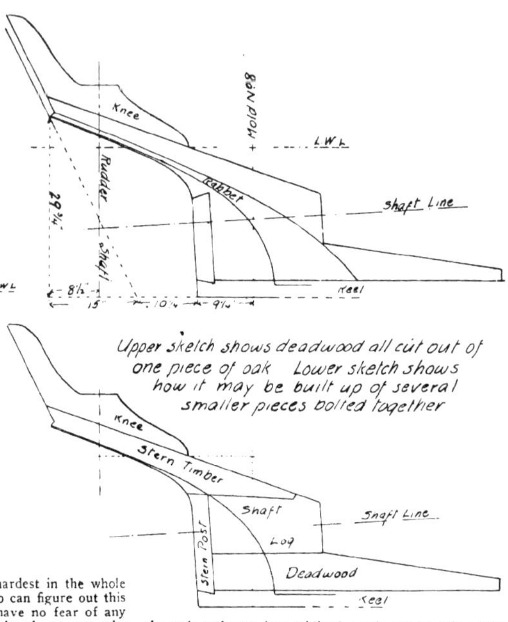

Upper sketch shows deadwood all cut out of one piece of oak. Lower sketch shows how it may be built up of several smaller pieces bolted together

This particular job is one of the hardest in the whole construction of *Sunfish.* Anyone who can figure out this transom and make a neat job need have no fear of any other part of the work. Apply the bevels as you take them, is the keynote.

The curve to the transom should be bent in it first. Steam the boards well in a steam box and then clamp them over a mold built for that purpose with about an inch more curve than you want; it will always straighten back a little, so put more curve in than you need.

Another way to build this transom is to bend about a ¾ or ⅞-inch transom and level the edges so that the plank ends go right past, flat-footed on its edges. then trim off these projecting plank ends to receive a ½-inch oak or mahogany facing-piece and bend and fit in this transom, fastening it with screws, plugged, to the inner rough transom. Be careful to set the transom up perfectly level when you bolt or rivet it to the knee that holds it to the deadwood.

When this is shored securely. and it is usually held by two stout timbers spread out like a pair of legs to the floor to hold it up and forward at the same time, you are ready to bend around the ribbands.

First run the sheer ribband—about a 2-inch square strip of clear spruce—in one length if you can get it; if not, join two pieces together by nailing a piece outside and lapping over the two ends where they butt. Don't attempt to scarph and rivet the two pieces of a batten together Sometimes another batten is bent outside of the first at the deck edge, as that is a very important part of the boat to keep absolutely fair and true. At intervals of about 6 inches run other ribbands fore and aft from bow to stern. They will, of course, be close together at the ends and once in a while the ends can be left out by using a shorter batten amidships.

Where there is a short curve in the frames, put the battens closely together, and where they are flatter spread the battens. Use square-headed coach or lag screws, turning them in with a monkey wrench after first drilling a hole so they will not split the wood and put a flat-iron washer under the head of each one so that you can pull the ribbands up wood to wood without having the bolt-head bury itself in the ribband.

She will begin to look very much like a boat when you get her this far along, and by standing off a way you can see just what her shape is going to look like. If, in putting on the ribbands, the molds do not seem to be fair, don't go and cut one mold to let the ribband in so it will touch the next, until you have carefully looked along the batten and tried your measurements. It may be all that is needed is to plane down the ribband a little, reduce it in size, or taper one end, and it may then bend in and still show a fair, easy curve.

No designer would think of trying to bend all the curves that make up a set of lines for a boat with one kind of a batten. They have many differently proportioned battens, some like your ribbands, all one size throughout their length; some larger in the middle than they are at each end; some large at one end and gradually diminishing all the way to the other end. So graduate your battens, if they are too stubborn, and don't blame the designer.

In the next number we shall start in and frame up the hull and proceed to plank her.

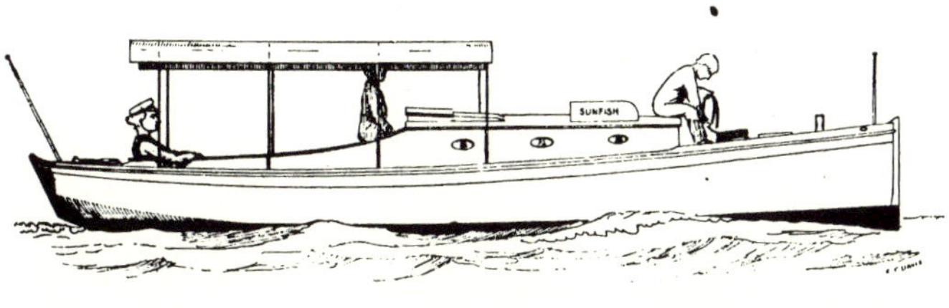

How To Build "Sunfish"

By C. G. Davis

Part II

WE knocked off work last issue with the molds and ribbands all in place ready for framing up *Sunfish.*

Before you proceed to frame her, see that all the seams that cross the rabbet line are fitted with what are called "stop-waters." With a half inch bitt bore so that it cuts half out of each side of the seam and drive in a white pine dowel so that any water attempting to flow up this seam and cause a leak will swell up this pine dowel and prevent the water from going through

You need steam and a steam box to bend the frames, of which there are forty pairs, each frame being about 7 feet long and 1½ by 1¼ inches, though there is an advantage in having them square, say 1½ inches, in that as you grab the hot frame from the steam box you can bend it on either of the four faces, whichever shows the most likely way to stand the strain. I have specified them deeper than they are wide for this reason The grain of the frame should be bent so that the plank fastenings go through the layers of wood and not through between the layers of the grain wedging apart, as it were, the layers of wood. If they are sawed out so that the grain would be across when cut on the narrow face you could never make a mistake in getting the grain right whichever one of the two narrow faces you bent against the ribbands

Many people like to bend the frames "on the flat" because they are easier to bend that way but for the good of the boat they should be bent on edge, as that is the way they have to resist the strains

An odd pair of gloves will be found very useful in handling the hot frames You need a number of 6 or 8-inch screw clamps and some one to hand you the frames from the steam-box Put the heel or lower end of the frame on the keel, your knee in the middle of the frame, and bend it just as if you were bending a bow to string it—bend it gently but steadily into place, and if you have a helper, which would be advisable, let him start to clamp the frame to the lower ribbands as you bend it down against them and follow on up to the turn of the bilge. If the frames are not steamed enough they will break, and if they are of poor quality they'll break anyway Rock elm makes a fine frame as it bends with very little breakage and is strong besides.

Space off along on the keel and ribbands where the frames are to go and mark with chalk so when you are working fast with a hot frame you can see just where each should go to have them evenly spaced. Hold the heads of the frames well in to give the round, tumble-home curve at the deck. They are liable to straighten back, anyway, as they cool off

When they have cooled you can nail them to the ribbands and remove the screw clamp to use elsewhere. Put the nails in slanting through the edge of the ribband into the face of the frame. Don't nail through the side of the frame from the inside of the boat into the ribband, for you will scar the sides of the frames.

By bending frames in hot this way you twist them with a monkey wrench so they lay flat ready to receive the planking without beveling them.

The heels of the frames should be cut so that they butt flat together at the center of the keel and have a slice taken off the under corner so that instead of the square corner of the frame touching the top of the keel they will fit flat on top of it out to the edge where the rabbet is beveled off to receive the edge of the garboard.

Drill a hole down through the frame and drive a 2-inch galvanized boat nail through into the keel at the heel of each frame. Then take some 1¼-inch oak boards 4 inches wide. Lay them on edge over the top of the keel and mark out the shape by running a pencil along the outside of the frames, marking this angle on the board. Then saw out this shape, or, clamping the board in a vise, rip it off with a draw knife and true it up with a plane. Fit the floors forward of amidships forward of the frames, and those aft aft of the frames. You can then bevel this floor off so that it gives an additional surface to which to nail the planking.

Rivet each floor to its frame with three round wire copper nails riveted over burrs on each side. Keep the upper edges of all these floors in a true line so that the keelson will not require much cutting and fitting when you run it fore and aft over them.

Away up in the ends where the frames make a sharp V, use wider boards to cut the floors out of and shape them down on the top or take an oak knee, slabbed up into 1¼-inch thickness, and get floors with a natural crook to them.

The keelson, a 3-inch square yellow pine stick, 24 feet long, is then bent down on top of these cross floors and held securely in place exactly over the keel until you bore holes with a long ½-inch auger bitt through keelson, floor and about 3 inches into the keel. Measure the exact lengths with a sliver of wood and cut corresponding lengths of ½-inch galvanized iron rod for drift bolts. Tap a slight head on one end by clamping it in a vise and using a ball pene hammer—a machinist's riveting hammer—to spread the metal. Then put a galvanized riveting ring over the end of it and drive the bolt home. One of these at each floor will hold her backbone solid as a rock, and if the keelson ends lap onto the deadwoods forward and aft and are bolted fast there the whole forms a very rigid truss.

The molds have to be removed to get in this keelson and the other long fore-and-aft stringers, but as you knock them out after unscrewing the fastenings into the battens put up temporary braces to the ceiling to replace those that

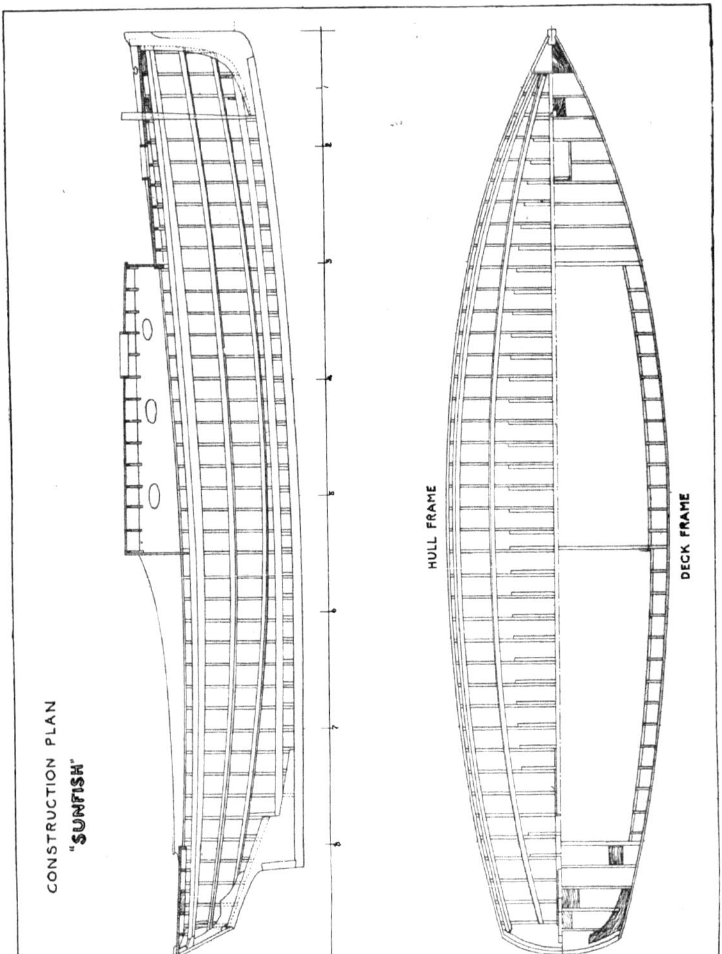
CONSTRUCTION PLAN
"SUNFISH"
HULL FRAME
DECK FRAME

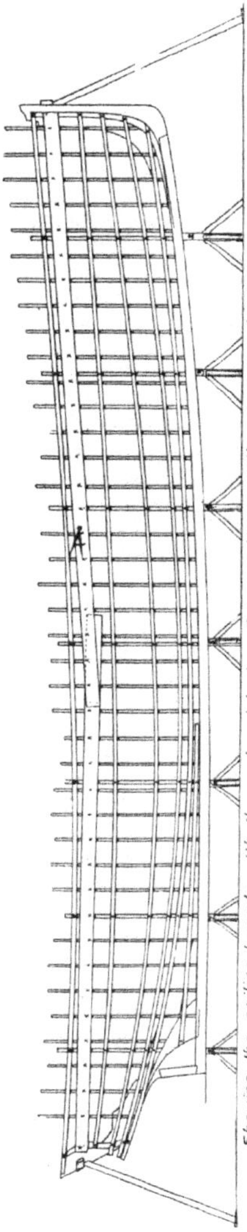

Showing the spiling boards with the spots pricked on it for marking out the shape of the top strake of planking.

held the molds, and nail cross braces to hold the frame apart to its proper width at each mold space.

Measure carefully just where the top edge of the 6 x 1½-inch yellow pine clamp is to come, which is the thickness of the deck ⅞-inch and the depth of the frame 1½-inch below the deck edge; 2⅜ inches in all. This clamp is the binder that holds the top ends of all the frames true into line. Although it is 6 inches deep in the middle, it should be tapered to about 4 inches in depth at the ends, and if you have a power planer handy, its thickness might be reduced to 1¼ or 1⅛ inch at the ends to help it bend, for it makes a pretty stubborn piece to handle. Steam it well before you try to bend it in place and you can then edge set it up or down, as you will find you have to, due to the tumble-home of the topsides, which point the ends down. It takes two pieces 32 feet long to get out these clamps, or one piece of 4 x 6-inch stuff ripped in two.

There are two bilge clamps on each side of 3 x 6-inch yellow pine, and though they may be a foot or so shorter, it would pay to order them all the same length, 32 feet, as the saw-mill would probably have to rip them all out of one big piece of yellow pine to get the length.

Use ⅜-inch galvanized carriage bolts to pull the clamp and bilge stringers snug to the frames at every other frame and then, when you bend the shelf in against the clamp, put a through bolt at every other frame clean through the whole lot, frame, clamp and shelf. This will give a stiff, rigid deck edge that will resist any bangs she may get alongside a dock.

Cut the forward ends of these clamps and stringers so that they butt flat up against the stern and transom.

Frame the deck before you start in to plank her up and it will be easier to work. The deck beams are all cut rounding with a "crown" or curve of 5 inches in 7 feet. Saw the deck beams out. Don't bend them. Bent ones have a way of flattening down again and there are not very many of them. Cut a wide thin board so that it forms a "she" pattern of this curve, and by trying this pattern at intervals along her deck you can make sure of getting all the short beams along the cabin space set true to the curve.

The deck beams are ⅞ x 1½-inch with an extra heavy one at the forward end of the cabin and at the after end of the cockpit where the deck ends. These beams should be about 1½ by 2 or 2¼ inches. Nail each of these deck beams fast to the clamp with a long, galvanized iron wire nail and under the inboard ends of them run a piece of spruce 3 inches deep by 2 inches thick fore and aft and nail them fast to it.

Go carefully over the frames with a batten before you start to plank and see that all the frames are true. Shave off a little here and there wherever a frame presents a hard edge, and when you are sure the frame is all true, start and plank her up.

It will take about 500 square feet of ¾-inch cedar boards to do this. Buy "dressed"—as planed-up lumber is called—and insist on good, clear lumber. You can't expect to get cedar without knots, but shun all sap, which is the bluish cast found near the edges inside the bark. Knots are solid but the sap turns to a soft punk that is apt to produce leaks. All small knots that show a black ring around them should be reamed out after the boat is planked and wooden plugs dipped in shellac driven in and sawed off flush.

There is no royal road to planking up a boat. I have had many people ask me if they can't make one pattern, and get a mill to saw them all out for them, that will plank up the whole boat. You cannot do this. Every plank requires a different shape, though the one pattern will, of course, do for both sides.

Take a "spiling," as boat builders call it, for the top strake. For this you want some very thin planks—about six in all—14 to 16 feet long, say 6 to 8 inches wide, and about ⅜-inch thick. Tack this spiling board lightly to the frame as nearly as it will go without being forced side-

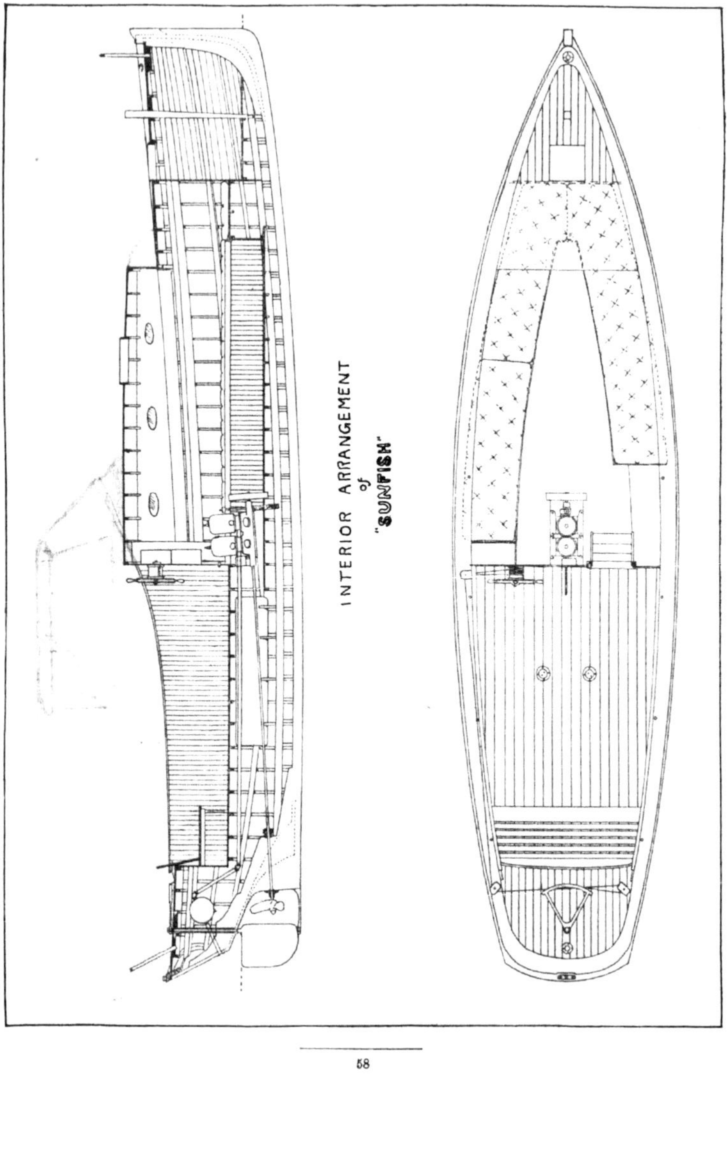

INTERIOR ARRANGEMENT of "SUNFISH"

ways and into the place where the topstrake is to fit. If it touches the sheer line at about mold number three it will be several inches too low at the ends. This board will give you the curve as far as amidships, aft tack another one the same way, and where the two lap amidships tack them together. Then, get a pair of carpenter's dividers. Set them to span the greatest space between the sheer line and this spiling board—screw the dividers so as to hold their legs apart—and from the sheer line at about every other frame prick off on this spiling board this distance; then by carefully removing these boards and laying them out flat on the plank, you are going to cut the top-strake out, if you can prick this distance back onto it and get the curve to cut the top edge so that when bent around the frames it will fit true along the sheer line. The lower edge of this plank is then marked out by bending a long thin batten so

Have the plank squeezed up good and hard to the frames with screw clamps, putting a chip under the foot of it so that it will not bruise the surface of the plank and rivet the plank on. Where the clamps will not permit riveting use flat-head brass screws 1¾ or 2 inch, No. 10. Scrape a little coarse brown washing soap onto the threads and the screws will turn in easier. Or if you can't afford screws use galvanized iron boat nails.

Use your spiling board again to find the shape of the top edge of the next two boards and in this way put on about three strakes of top planking.

As the top plank is a sort of binder, many prefer to make that board of quartered oak or of yellow pine. Yellow pine is good and you easily can get that kind of wood in lengths long enough to make it all in one piece.

Then spile in the same manner for the shape of the gar-

Section at Mold No-6.

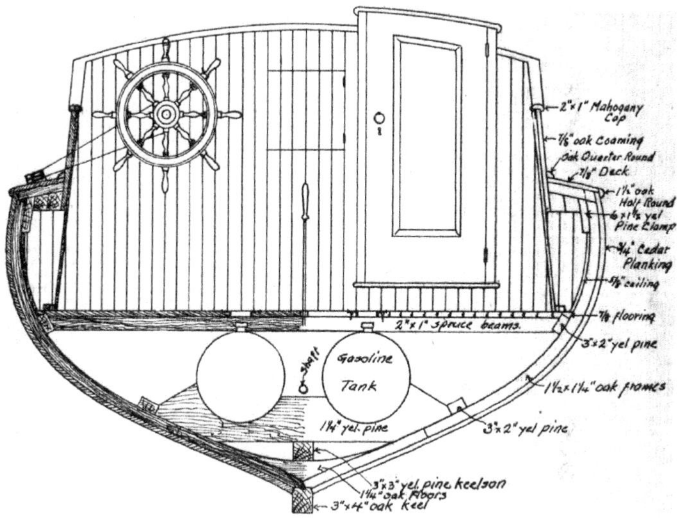

that it makes a fair curve, leaving the plank about 4 inches wide amidships and tapered to about 2½ inches forward and 2 inches aft.

In putting this top strake on, have two braces. In one have a bitt to bore for the wooden plug, about a ⅜-inch auger bitt if your 2½-inch copper wire nail heads will go into that sized hole without tearing the wood; if not use a 7/16 inch, or even a ½-inch bitt. In the other brace have a gimlet bitt—a breast drill with a small bitt works faster and easier and is more generally used by boat builders—some shops having an electric drill that goes through the wood as if it were cheese, and is a great time-saver. Follow through with this smaller bitt, boring a hole into which the copper nail squeezes tightly

board where it fits along the keel. Cut your thin spiling boards so that they fit roughly to the shape at the ends and get out the garboards. They, too, should be of oak or yellow pine, but instead of being wider in the middle than at the ends, they are just the reverse. The idea is to fill up the surface with the top strakes and the garboards so that the remaining space to plank up will be more like a barrel and take boards more of an even shape and size. If you cannot do this with the garboard alone, put one or two more strakes above it, the first and second broad strakes as they are called, making them about 6 inches wide and tapered so that the rest of the space can be divided equally both at middle and ends.

To keep the intermediate planks of uniform size and to

give fair planking lines, get out the rest of the planks by using a "planking scale." This is nothing but a thin slat of wood, thin enough to bend easily around the frames so scaled off that you can measure how wide each plank must be at any frame to fill up the space between the top-strakes and broad strake already on. Say it is 48 inches at the

Section at Mold No. 5

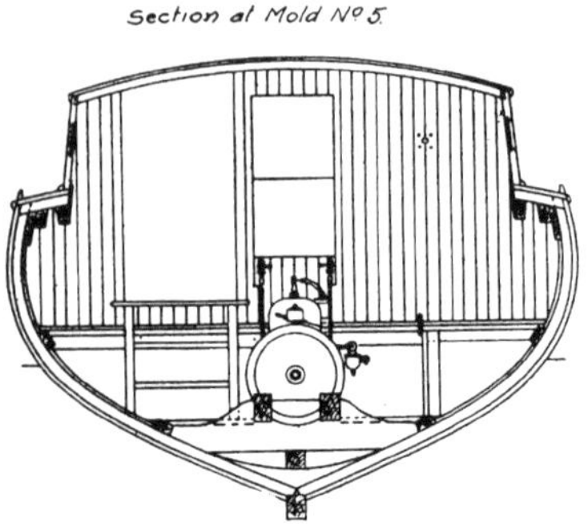

nearest frame to amidships and we want the plank 4 inches wide. It will therefore take twelve such planks to fill that space. Bend the batten with one end against the edge of the broad strake around the frame and mark where the edge of the last upper strake comes. Mark that "4."

Then do the same at Mold No. 1. There twelve planks would make each one only 3 inches wide. Bend the batten here and mark it "3." Divide the distance between 3 and 4 on the batten or plank scale into eight equal parts to represent eighths of an inch. By bending this planking scale at any frame this will show you how wide the plank should be at that frame.

You cannot expect to have the planks all in one length from end to end but will find that you have to make them of two pieces. Make the joint, or "butt" as it is technically termed, midway between the frames and rivet the ends to an oak butt block about an inch thick and wider than the plank so that its edges hook behind the plank above and below it. The last plank, called the "shutter," must be spiled for on both edges and must be fitted very carefully.

Plugs can be purchased to fill the holes over the nail-heads after you have riveted up the copper nails that hold the planks on. Dip each plug into either glue or thick white lead paint and tap them lightly into the holes, snipping them off flush with the plank with a big chisel. only learn to snip them so that the grain of the plug doesn't go in and make a hole.

With a plane jack down the seams a little and then with a roll of spun wicking, a calking iron and a mallet, calk each seam. Drive the cotton in so that it is at least an eighth of an inch back from the edge and smear it with a mixture of thin white lead applied with a seam-brush—a single row of bristles set in a wooden handle.

The planksheers can be bent after steaming well around stout cleats nailed to the floor and fitted when cold, or they can be sawed out in sections and scarphed together.

To make the mold to bend it, simply drop a plumb line down from the deck edge at intervals and nail the cleats to these spots. This way saves a lot of unnecessary measurements. Before the planksheer is fastened down toe-nail in a row of blocks between the beams half under the planksheer and half projecting to land the ends of the deck on. Plug fasten planksheer, using galvanized iron wire nails.

The deck is laid in strips of white pine 3 inches wide and ⅞-inch thick. Do not try to make it absolutely tight, at the joints, but run a thin shaving off the upper half of one edge to leave an open seam on top while the bottom is jammed tightly together. Then calk these seams and "pay" them with white lead paint, the same as the planking was treated.

Where the tiller rope sheaves are to be, fit an oak block under the deck between the beams and the same forward where the hatch and the mooring post come.

If you can get your motor, now is the time to build the bed for it and line it up through the shaft hole, fitting on the stern bearing outside and the inside stuffing box. Cut saddles to hold the gasolene tanks securely in the bottom of the boat. Pipe your exhaust out through the stern and then lay the cabin and cockpit floor beams and floors.

The cabin floor boards can be 6 inches or so in width, but lay the cockpit floor in 4-inch strips just as the deck was put down.

Square up the inner edge of the deck around both cabin and cockpit, and around on the cockpit floor directly under this edge fit a rabbeted white pine sill piece to set the lower ends of the tongue and groove chamfered staving into. Let the upper ends stick up high enough above the deck to form the shape of coaming shown in the plans.

The cabin sides, 1¼ inches thick, are landed flat on the deck after smoothing off and putting them where it is to set. If you cannot get one wide board, build it up in three strips with upright ribs inside to hold them rigid. Drift bolt the cabin through from underneath.

The cabin beams are ¾ x 1¼-inch with a crown of 6 inches in 6 feet notched into the cabin sides and nailed. Over them is laid a ⅝-inch wooden deck, and after this is given a coat of paint, a thin piece of canvas is stretched tightly over it and tacked over the edges with copper or galvanized tacks, which are all hidden under an oak half-round molding.

The after bulkhead is staved up and down in the same stock as the cockpit staving. The outer thickness of the cockpit coaming is then fitted up against the after end of the cabin sides, its after end shaped down in a curve to the deck, and an oak cap rail fitted over the two.

After the deck is all planed, sandpapered and varnished, one coat, fit the low thumb rail around the edge of it, plug fastening it with galvanized boat nails to the planksheer, and put on the oak half round moldings at the deck edge.

Sandpaper the hull outside, putty all the seams and holes, and give it a prime coat of paint. When this dries scratch in the waterline or paint line if you want her to show a boot-top, and give another coat of white paint above water and copper paint below.

The stem band you can make out of a strip of narrow half-round brass. the skeg, rudder and quadrant can be bought, and a rudder port made by screwing a stout piece of brass pipe, just large enough to take the 1¼-inch diameter bronze rudder stock into the hole through the overhang and sawing it off flush with the deck, or about ⅛-inch above it.

The flagpole sockets, windows in the house and stern seat are fittings that anyone who has gone thus far will need no instructions for. There is the steering wheel and sheaves, hatch in the cabin top and cabin doors to finish, and the interior we leave for every man to fit up as his fancy dictates. By extending the transoms forward 3 feet under the deck, there is transom accommodations for two people to sleep to port and one on the starboard side well clear of the companionway.

Editor's Note.—Detailed Instructions for Planking, Calking, Painting, Laying Canvas on the Cabin Top, Building a Motor Bed, etc., will be found in Motorboat Handbook, Volumes I and II. Readers who wish to build *Sunfish* will find the Handbook a practical aid in the work. Those who do not own copies of the Handbooks, may obtain them by sending a dollar and a half, which is the price for both volumes, postpaid.

HOW "MOLLYHAWK" WILL APPEAR AFLOAT

How to Build "Mollyhawk"

Part I

By C. G. Davis

A MOLLYHAWK is not a bird of beauty, therefore the name is appropriate in this case, but beauty is not always the most desirable element in a boat, and if our *Mollyhawk* shows the staying qualities, the seaworthiness and lasting qualities of that homely sea fowl so familiar to deep water sailors as to be part of their everyday life, she'll be as good a boat as any man who loves the life on the water will care to own.

She is a boat that will go through stormy weather as a motorboat should go, easily and dry, not floundering about in the surface water. This desirable quality has been obtained by making her more of a boat, a deeper boat and a heavier constructed boat, than the ordinary 28-foot craft.

She looks, in profile, somewhat like a small edition of a steamship—her stern will stand the slopping of seas without the jar felt in flat-sterned boats, and her bow is high enough to take care of any comber she may be put to. Her deckhouses look a little peculiar, due to the long, narrow skylight, which, while it looks odd in a side view, does not show up at all badly in the real boat, and its advantages are twofold. It gives full headroom in a boat that would otherwise be all out of proportion, if the cabin itself were carried up that high, and in Summertime makes a nice cool cabin by the ventilation possible with such a cabin construction. In the flat paper plan you see the full height, but in the real boat the perspective sets it back, and it appears considerably lower.

Mollyhawk is 28 feet long, has 8 feet beam, and draws 2 feet 6 inches of water, with a good, heavy displacement, 9,920 pounds. This heavy displacement is intentional, as a study of the lines will show. She is big below water, and it will require some concrete ballast to bring her down to her designed waterline, as the weight of her hull engine and fittings will hardly equal 9,920 pounds; more likely they will be about 7,000 pounds, leaving about 2,900 pounds of ballast.

If you have been out in a light displacement boat in heavy weather, you know how they throw you about, and, when I say heavy weather, I don't mean a Summer thunder squall on an inland lake, where it's all spray and no sea, but take the fleet of boats that go out down New York's lower bay, or the style of boats that navigate out of 'Frisco Harbor, and go out into the ocean fishing, and the conditions such boats have to contend with are what this boat is designed for. Another condition she is adapted for is where a man wants a boat to take long cruises and live for months aboard his boat. He can live on such a boat as this in comfort, and be able to walk when he gets ashore without having to get the kinks out of his legs, or feeling the ground heave and pitch under him.

Don't start to build *Mollyhawk* unless you have first tried your hand at some smaller craft. Not but what you might succeed—that is a matter of individual ability—but what I mean is that *Mollyhawk* is not a primer, it's one step higher, is a second-grade reader so to speak, a little more difficult and much more of a boat.

I have laid out all the work, and the plans herewith published show how the work is to be done, but it's up to you to open your tool chest, go see the lumber dealer, and prepare to use your muscles. You who want to keep your waistband girth down can leave off your dumb-bell exercises and your walks; the contortions you will go through in building a boat are the best kind of exercises to keep a man healthy and strong.

Build her under cover, if you can, inside a barn or shed. If this is not available, and I never was so lucky myself, do as I did, set up some posts and from these run rafters up against the side of your house, and roof over and board in with novelty siding.

The first step in the real work of construction consists in getting out the molds. These molds are wooden patterns, as it were, that show the boat's shape at the various stations marked in our plans, 1, 2, 3, 4, 5, 6, 7 and 8. No 8 could be dispensed with, but, on account of the boat being rounded up so quickly at the stern, I have shown it, and it will pay you to go to the trouble to make it, you being an amateur. These molds are thrown away when the boat is completed, and for that reason are generally made of some very cheap wood about one inch thick. Their shapes can be laid out from the measurements given at six-inch intervals in our plans on Plate 2. When these are all ready, lay them aside

Plate. 1.

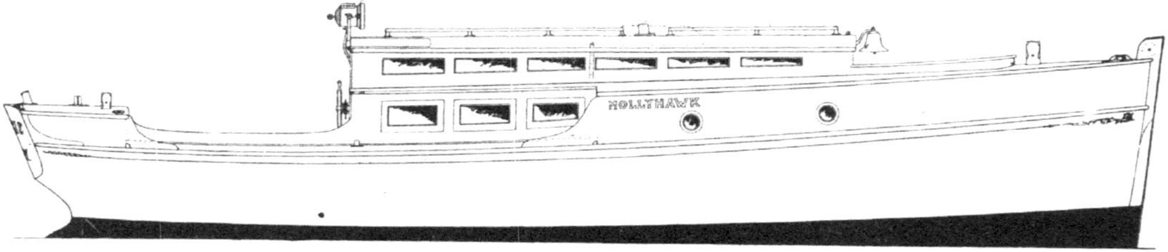

SIDE VIEW OF 'MOLLYHAWK'

DECK VIEW

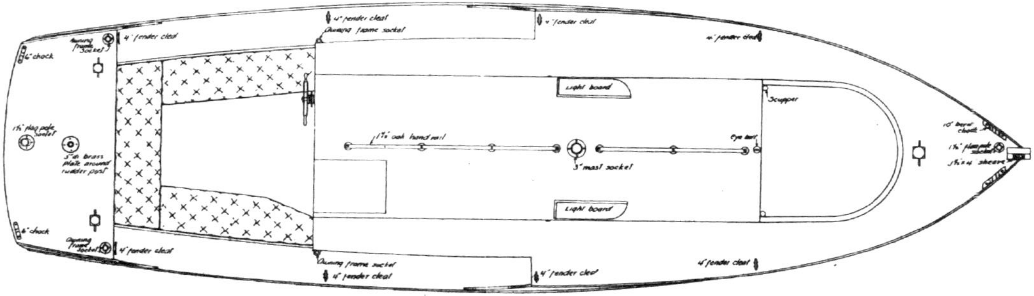

Lines of 28 ft. Cruiser MOLLYHAWK

Designed by C G Davis

Plate 4.

DIMENSIONS

LENGTH OVER ALL	28 ft · 0 in
LENGTH ON WATER LINE	25 ft · 6 in
BEAM EXTREME	8 ft 3 in
DRAFT	2 ft 6 in
DISPLACEMENT	9920 lbs.

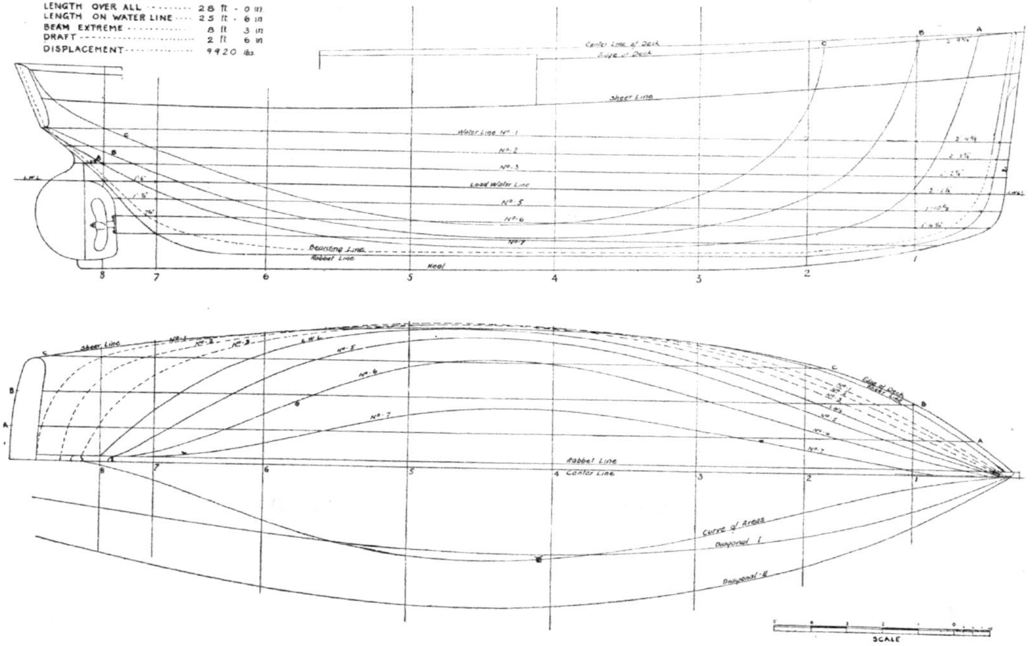

and start on the oak keel and the backbone of the boat. The various members which constitute the backbone are all shown and identified in Plate 3. Nearly all the members here shown are to be of oak. Sometimes the knees, such as the transom knee, stem knee, etc., are made of hackmatack. Either wood will do, but by all means make your keel, shaft log and fore foot of good, sound, clear white oak. The keel is the first and largest stick in the boat. It is, as shown in Plate 3, 22 feet 1½ inches long, 4 inches thick and 6 inches deep. Don't make the mistake so often made by amateurs of ordering a 4x6 of your lumber dealer, and then, when he asks if you want it rough or dressed lumber, tell him you want the

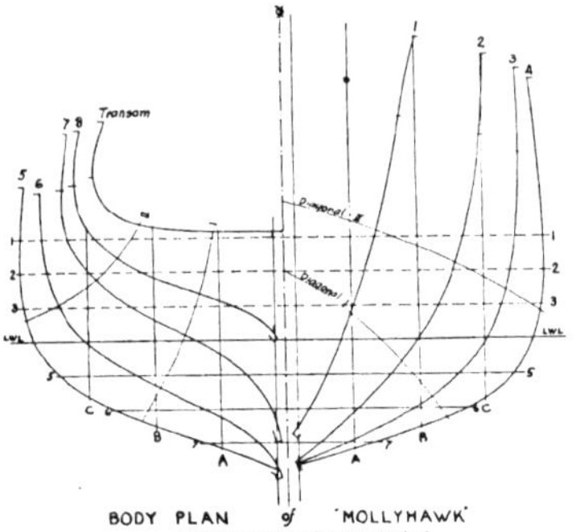

BODY PLAN of 'MOLLYHAWK'
Moulds are drawn to inside of plank

wood dressed, for, if you do, you will probably get about a 3¾-inch stick for your keel, and, while this at first thought might not seem to make any difference, you will find it, as well as the rabbet line, has all been figured out for a 4-inch stick of wood, and this would all be changed with the thinner size. There is very little cutting to be done on the keel in this particular kind of a boat. The general run of boats that are built nowadays use a flat plank for a keel and bend it up aft, so that it comes right up to the transom, but this is no light runabout. We want weight, strength and rigidity in this boat, and for that reason have gone back to the regular old ship style of construction. The deadwood and shaft-log shou'd next be cut to the various shapes shown in the plans, and, by being made in two pieces, the shaft-hole can be planed or gouged half out of each, instead of being bored with an auger, or, if you prefer, the deadwood can be in one piece and bored, but ninety-nine out of a hundred amateurs would find great difficulty in getting an auger large enough to bore a two-inch hole. The stern timber has more shaping to it than any other piece of wood in the boat, and, if the amateur could make a wooden pattern of this and have it sawed out at the saw mill where he ordered his lumber, it would save him considerable cutting. If not, he will have to line it up and saw and chop.

The forward end of the keel is built up with a four-inch fore foot, which takes a stick of timber 6 feet 3 inches long and 9 inches wide, a stem which is 6 feet 6 inches long, 4 inches thick and 9 inches wide. These are all shaped and held together by the four-inch hackmatack knee to which each is riveted. The bolting together of this deadwood is generally a sticker for the amateur, but if he will provide himself with two or three half-inch galvanized iron rods—they generally come in standard lengths of about 12 to 14 feet—about four pounds of half-inch rivet rings, and a long ship auger, also half-inch, he will find the problem to be greatly simplified. Carefully set the various pieces of wood one on top of the other, tacking them lightly with wire nails, then bore the various holes where you want bolts; with a hack saw cut the iron rods into the proper lengths for bolts to go through these holes. Then clamp these bolts in a vise, and with a riveting hammer head up, as it is called; that is, burr over one end so it makes a head on the end of the iron. Slip one of the rivet rings over this, and drive it through the hole from underneath, where you should previously have bored a hole in about one inch, large enough to take the riveting ring, and, if you cannot get an auger large enough, cut it out with a gouge. This is to countersink the rings so that they will show flush with the wood or so a wooden plug can be put in to fill the hole. A bit of tallow rubbed on the iron bolt before you start to drive it will save a good many blows of your top maul, as the light sledge hammers are called, when you come to drive these bolts through the auger holes. They should be just long enough to stick up about a half inch above the inside edge of the wood. Then, with somebody holding a heavy sledge hammer or other heavy weight against the bolt on the under side, slip a rivet ring over the bolt and proceed to head up the inner end. It is rather difficult to tell an amateur just when he has headed the bolt up tight enough. An experienced man could tell by the sound of the blow on the bolt. The only way I can explain to you to tell when it is sufficiently drawn up is to tell you to keep tapping the bolt and burring it over until the rivet ring begins to sink into the wood. The deadwood bolts need not necessarily go clear through. They can simply be driven through the keel and into this deadwood, a distance of about 6 inches, forming what are called blind bolts. If all these bolts were put in exactly parallel to each other, you can readily see that it would be an easy matter for the wood to draw away again, but, on the other hand, if the bolts are at different angles, staggered, as boat-builders call it, the wood is firmly locked and cannot get away.

Before you put the various pieces of wood together, paint their surfaces with a good thick coat of white lead paint, any color will do. Shipyards always have a great many screw clamps of various lengths, and it would be well for you to provide yourself with them, as you find you need them for this job, as, for instance, in pulling these deadwoods together you will find screw clamps a very great help. Of course, an ingenious mechanic can overcome it by making a dog, that is, cut a notch in a piece of oak a little larger than the span of the wood to be bolted together, and wedging them together by inserting a wooden wedge between the upper part of this dog and the top of the deadwood, and driving the wedge in solid, but dogs will have to be put on each side of the deadwood to prevent tipping to one side. This will squeeze the wood together about as tight as it could be drawn with a screw clamp, and you are then ready to rivet.

This backbone is generally laid on its side while being bolted together, and, while it is still in this position, it is a good time to cut your rabbet. Either lay it out flat on the floor, being careful not to strain any of the joints, or, better yet, lay it over wooden horses, which always come in handy around a boat shop, and which it will pay you to take the time to make before you start. This raises your work off the ground at a convenient height, and keeps you from almost breaking your back by trying to sit on the ground and hammer. But if you're going to cut your rabbet at this stage of the game you will have to be very particular about getting just the proper bevels at which to apply a little sample of your

Plate. 2.

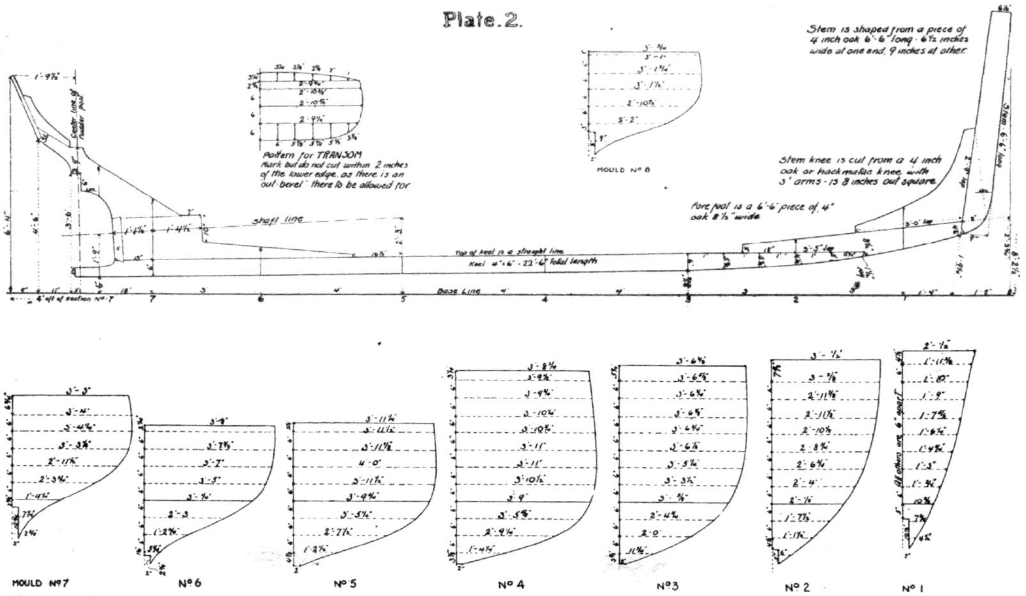

Patterns for the temporary moulds, which when set up in their proper places along the keel give a good idea of the MOLLYMAWK's shape. Battens are then bent from mould to mould, and over these battens the steam softened frames or ribs are bent.

planking when you cut the notch forming the rabbet, and I would hardly advise doing so unless you want to go to the trouble of laying down all the boat's lines just as a regular boat-builder does. If you do this, you know exactly the bevel at which each waterline intersects the stem forward and the deadwood aft, and this gives you the bevel at which to apply your pieces of planking. But following the manner in which we are building this boat you had better leave this rabbet line until after the molds are in shape, when you will be sure of making no mistake, as you can bend battens around these molds and then cut your rabbet line so the end of the batten makes a perfect fit.

It would be well to give the keel a coat of lead colored paint, to prevent the wood drying and checking or cracking, as the keel will probably be set up several weeks or months before it is closed in with planking, and all this time the wood is exposed to the action of the sun and wind. Set the keel up so its lower edge is about two feet off the ground, or eighteen inches at least. If you get it any closer to the ground than this, you will experience great difficulty in putting on the lower planking. The angle at which the keel is to be set up is shown in Plate No. 2. It is slightly higher at the forward end than it is aft. If you are building this boat in a shed, where you can shore the fore and aft ends of this deadwood to the rafters overhead, you will find it will be much easier to work around your boat than if the shores have to be braced up from the floor or ground upon which you are working. You will always be stumbling over them, and, when you come to set the molds up at the various distances also shown on Plate No. 3, shore them overhead for the same reason, if it is possible. If you have to shore them to the ground, dig down far enough so that you will be sure the ground you shore them to will not shift, as the least shifting out of line of these molds will throw everything unfair, and cause you a great deal of annoyance after. Brace them in a fore and aft direction as well as sideways, for you have to put considerable strain on these molds when you bend your ribbands, and later when you bend the frames over the ribbands.

Another point to be remembered in setting up your molds is to put them so the after face of the forward molds and the forward face of the after molds come just at the mold station marked on the keel. The reason for this is the bevelling away from the amidships towards each end. The shape you want is the shape of the face that is even with the mark on your keel. They are the working faces, and, when you run the ribbands, bevel off the molds so the ribbands will fit flush with this edge. The transom will have to be made, as it would be foolish to make a temporary mold of the shape of the transom, and then to have to make a new one, the real transom itself, later. Get it out to the shape shown out of 1¼-inch oak which has previously been steamed and bent so it has four inches round to its face. This transom can hardly be gotten out of one piece of oak, as it is over two feet wide. It would be more desirable if you could do so, if you cannot, bend the pieces of oak over the mold which gives them the four-inch bend, and then dowel their edges together with wooden dowels; mark the shape of the face as shown, but, when you cut it out, do not cut within two inches of the lower edge. The reason for this you will soon see when you bolt this transom fast to the little hackmatack knee, which holds it to the stern timber. If you do cut it away you will then see, owing to the bevel at which the ribbands go around the No. 8 mold, you will have to cut away still more on the transom to make the ribbands fit flat against it. By having previously left this wood standing, you can now do your cutting so the ribbands will finish fair with the desired edge.

While you are putting up the molds, let the lumber yard be getting out for you sixteen strips of yellow pine, 2x2, about 30 to 32 feet long, in one piece if possible. If not, they will have to be spliced out of shorter stuff. You will find that most of these you can bend around, starting at the rabbet on the stem, following the sheerline for the upper one around the molds to the transom. There may be a few, however, where the two-inch stringer will be too stubborn to bend around the quick curve at the stern; if so, plane them down, until they can be bent around, but don't make them too easy; they want to be just as stiff as they can possibly be put around, because they have to hold the frames when you come to steam bend them and make them take the proper bend, and, if these battens are not very stiff, the frames, when you come to bend them, will pull the battens out of line, and make an unfair surface to your boat's side. It will require about six of these battens on each side of your boat, and, at the turn of the bilge from amidships aft, you may even have to put in another ribband, so as to make the frames take a true bend, and not straighten out from ribband to ribband when you come to bend them. Fasten these ribbands at each mold with a good stout lag screw, say 4 inches by ¼ inch, turned in good and strong with a monkey wrench. You can form a very good idea of the shape of your boat by this time, as her shape will be very clearly outlined by all these fore and aft ribbands. Most amateurs feel very much elated when they arrive at this point, but all this is superstructure, and is all to be torn away as the real boat progresses, which is the case from this point on, for you are now ready to begin and bend in the real timbers of the boat.

The frames are to be 1½ inches square, spaced ten inches apart, so, while you are waiting for the clear, sound white oak to come from the mill, for it will pay you to

Plate 3.

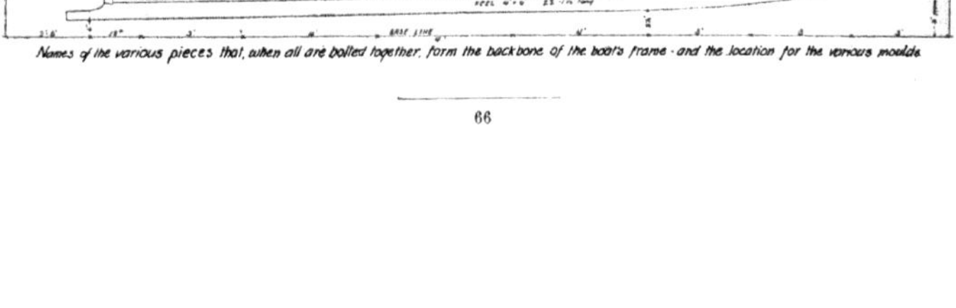

Names of the various pieces that, when all are bolted together, form the backbone of the boat's frame - and the location for the various moulds.

order all this from the lumber yard, cut up and dressed to 1½ inches in 8 to 12-foot lengths, take a thin batten of pine, about a quarter of an inch thick and inch or inch and a half wide, space off ten-inch intervals along your keel, and from these bend the batten up around the inside of the ribbands, and mark with pencil this location. When you come to bend in the hot timbers, you can bend them right along these pencil marks. This will insure the frames being evenly spaced, and not all standing zigzag. Of course, you can straighten the timbers up after they are cold, but along in the ends it will save a whole lot of trouble, on account of the excessive bevels there if you don't have to shift them. To soften your timbers so as to make them pliable enough to bend in around the ribbands, you will have to build a steam box. This subject is one which has been fully explained before in MOTOR BOAT, and along with many other subjects that will enlighten one in building this boat may be obtained by getting copies of MOTORBOAT HANDBOOK, Vols. I and II. It takes generally about half an hour to properly soften the frames. Cut the heel of the timber roughly, so that it fits flat against the keel as you bend it in. Two can work to far better advantage than one at this job. A 1½-inch timber, even when saturated with steam, is a pretty stubborn piece of wood to bend, and a day's work at this will tire any man. Let one man get inside the boat to work, the other handing him the hot timbers from the steam box, then, as he bears down with his feet to crimp the frame down into its place against the battens, pulling the head of the frame inboard at the same time, his friend on the outside can follow along, starting at the keel and clamp the frames to each one of the ribbands in succession. These clamps should not be removed for five or ten minutes. At the end of that time, after nailing them off with about two-inch wire nails, driven diagonally through the ribbands into the frames, these clamps may all be removed and used farther along on the hull. Very little difficulty will be experienced in bending the frames from the bow clear back to about section No. 5. From there back you have to be a little more careful on account of the reverse bend in the heels of the frames, and the bend at the bilge becoming more sudden. There is a way of easing this part of the operation, and that is to split the frame down as far as where the quick bend comes. This allows the inner half to slide on the outer half, and yet, when the plank fastenings are put through, they rivet the two securely together again. Personally I would not recommend this kind of construction.

After the frames are all in, sawed oak floors of two-inch oak are fitted alongside the heels of each pair of frames, riveted to them, and securely bolted to the keel with about 1½-inch iron. The shapes of these are very easily determined by laying a thin pine board across the top of the keel and against the frames where the floor is to fit, and marking the shape of the outer side of these frames with lead pencil on this board. With a few cuts of a draw knife you can cut this thin pine pattern and use that as a templet to mark out the shape of the desired floor. If you could only build a boat as easily as you can tell how to do it, we could build about one a day, but when the amateur tackles this job of fitting in two-inch oak floors, unless he has a band saw handy to help him, he will do very well if he fits in from four to six of these floors alone in a day.

How to Build "Mollyhawk"

Part II

By C. G. Davis

WHEN the heels of the frames are all secured by the 2-inch oak floors to the keel and deadwood, get out the fore and aft clamps from 2 by 4 inch yellow pine stock. There are eight of these in all, the upper two being short ones. Be very particular in putting in the main deck clamp, as the deck line will show unfair if this is not done, and no one wants a boat with a wavy, snake-like sheer. No matter how carefully a man may design the boat or how accurately measurements may be taken, it is always up to the man who builds the boat to see that the lines run fair and true, and this can only be done by sighting along the side of the boat standing at one end; while another man, with a hammer, taps the clamp up or down to take the kinks out of it. It is all a question of accuracy of a man's eye in being able to spot these unevenesses.

The two lower clamps are the same size as the upper ones, if you want to you can reduce them in width and thickness toward each end. This is generally done by the careful professional builder, but nine out of ten amateurs, not having machinery at hand, seldom go to the trouble of planing down these two-by-fours.

Diagonals Numbers 1 and 2 in the body plan of *Mollyhawk* show about the position for these two lower clamps. If run in about this line they will practically lie flat on the face of the timber or will be so nearly so that they can easily be screwed down with screw-clamps, and will not require previous shaping. All these clamps are to be bolted to their frames with 4-inch by 5-16 inch carriage bolts. The head of the bolt should be let in flush with the face of the frame and a washer put under the nut on the inside. This washer will permit your turning the nut up good and hard, drawing frame and clamp tightly together. Do not put all the bolts along in the center of the clamp but stagger them so that one will be near one edge, the next near the other and so on, alternating them. In the end you may have to favor one side more than the other to pull the twist out of the clamp.

Before these clamps can be put in it will be necessary to remove the molds, but be sure and tie the boat up from side to side with some good stout stay laths before you do so. Otherwise the hull may sprawl, as boatbuilders term spreading.

At this stage of the game most amateurs are in a great haste to start the planking of their boat. They want to see it grow, but they will get ahead far faster if they will leave the planking alone until they have the entire deck frame complete. They will find it much easier to do this part of the work. Everything is so much more accessible and there is a better light to work by than will be the case when the hull is all shut in.

As we have designed this boat there are no very long deck beams, the longest being the three across the stern. All the others are short on account of the midship skylight, but it would be well to get out three or four long ones, long enough to go across the entire boat, to hold the superstructure in shape while you are fitting the shorter beams, and then saw them off later.

As all these deck beams are exposed when the boat is completed, see that they are nicely smoothed up and, if you want a still fancier job, champher the lower edges before you put them in. Fit the deck beams in alongside of the head of the frames and fasten their outer ends to the clamp on which they rest with 3½-inch heavy galvanized iron nails, with the heads sunk in so that when you come to plane and dress off their tops so the deck lies perfectly true on them you will not be striking the plane on a nail-head. All these main deck beams should be marked out from a pine templet made with a crown or curve of 3½ inches in 8 feet. Do not, above all things, try to lay a flat deck as many amateurs attempt to do. There is nothing shipshape or "boaty" in a flat deck and there is a very good reason why it should be crowned. That reason is, to make it shed water.

There are about four corners that should be reinforced by fitting and bolting in oak or hackmatack hanging knees about 1¼ to 1½ inches thick. This means eight knees. They should be located one each side where the forward beam of the after deck comes, one each side of the after end of the main cabin, one each side of the after raised deck and one fastened at each end of the deck beam that goes clear across alongside the forward end of the cabin. Rivet these to the beams with at least two ¼-inch bolts to each arm of the knee. These may save a whole lot of straining when your boat takes a side bump against a dock and they are well worth the time it takes to put them in. For the same reason large knees are fitted in the corners where the transom joins the side planking and a breast hook is put up in the bow.

As you cannot lay any more beams until you have laid the two little pieces of side deck and put up the cabin sides which carry those beams, they can be left and the more interesting work of planking up the hull taken up.

Planking up a boat may seem simple and to some it turns out to be such a job, but if one has never considered this subject carefully and examined other boats so that he has a fair idea of the way in which it is done, he will find himself working into all kinds of difficulties. Old subscribers to Motor Boat will have some understanding of this subject from the practical articles printed. Those who have not read these articles will be able to secure them all in compact form in the Motorboat Handbook. Not only is the subject of planking explained therein, but the bending of frames, deck construction and many other valuable articles both on boat construction and the use and the care of the boat after she is built.

Briefly stated, the process of planking may be compared to the construction of a barrel. As the barrel staves are made wide in the middle and narrow at the ends, so is the yacht planking wider amidships. Just how wide each plank shall be you have to determine by bending a thin batten around one of the midship frames and dividing it up into such widths as your stock of planking will permit. For instance, if most of your cedar planking which the lumber yard has delivered to you will only allow you to get out a plank five inches wide in the middle without leaving bark on the edge, do not lay your boat for six inch planking. This distance, measured on the batten in inches, divided by five, will show you how many planks your boat will require. How wide these planks will be forward or aft at any other frame can be determined the same way, by dividing the length along the frame to be planked into the number of planks that are being put on amidships, which would probably give you something like 2¾ or 3 inches.

The top strake, or "sheer strake" as it is called, and two or three more below it and then the garboard or plank which goes next to the keel are generally fitted in first and then the space between is divided up as previously described. To find the shape of the garboard requires what

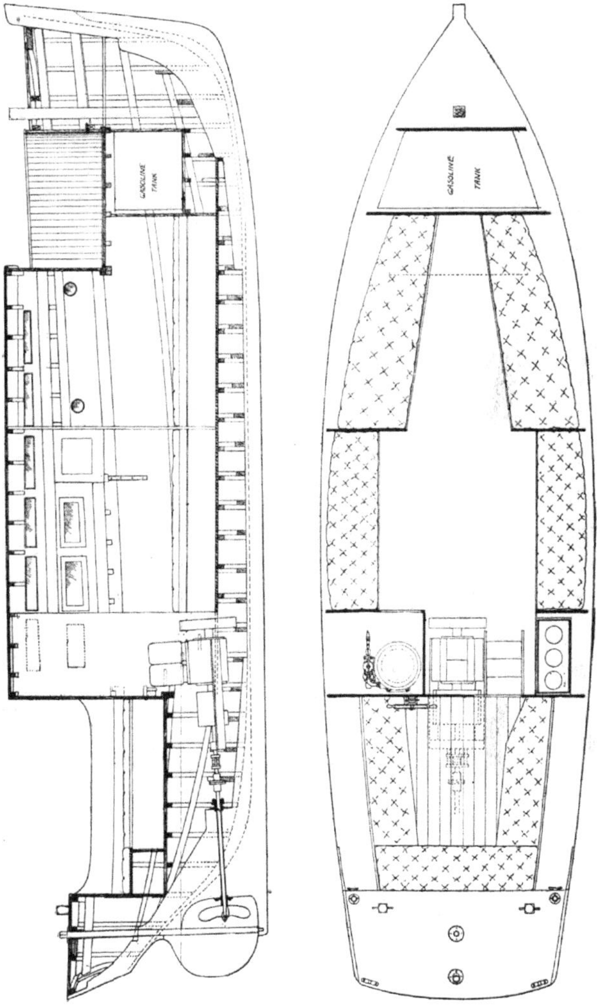

ARRANGEMENT PLANS OF "MOLLYHAWK"

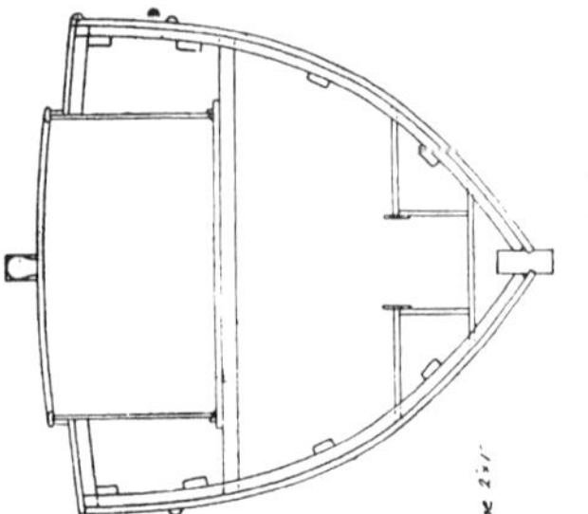

SECTION · NO · 2
shows forward cockpit

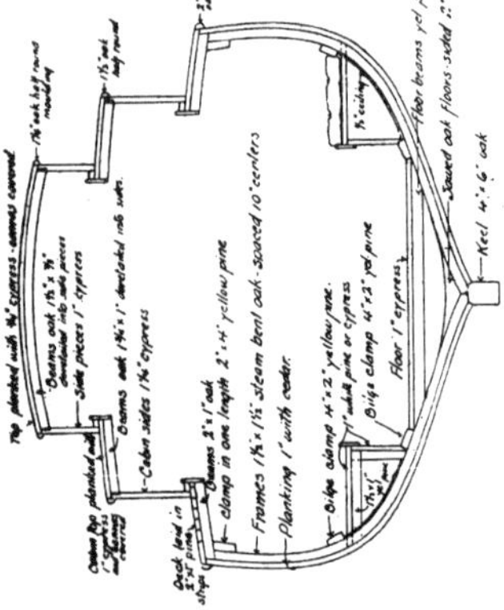

SECTION · NO · 5
Giving sizes of various parts

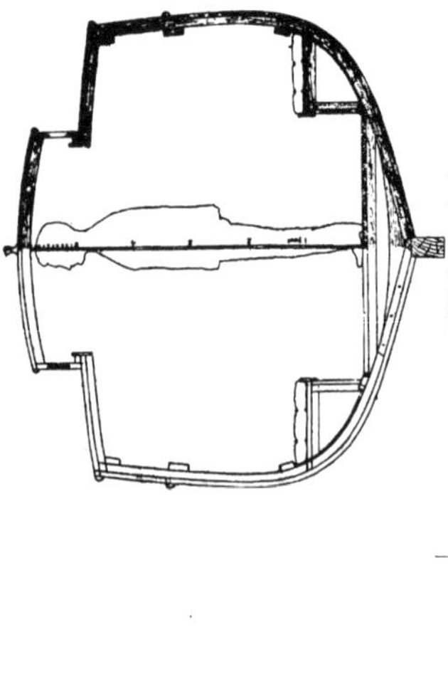

SECTION · NO · 4
showing the comparative height of a 5'-9" man

is called "spiling." In other words, it requires the spoiling of one plank which is generally a thin pine or cedar board about ⅛ of an inch thick. This thin board is used as a pattern cut roughly by eye so as to fit along the keel, and then, with a pair of compasses, set so as to span the greatest interval between the edge of the rabbet and the edge of this pattern, proceed to prick off a line of spots along the pattern, keeping one point of the compasses at the edge of the rabbet. By laying this pattern out flat on the one inch cedar board from which you are going to cut your garboard, and pricking these distances back you can readily see that you get a line of spots the same shape as the rabbet against which the edge of this plank must fit. This "spiling" process is repeated for almost every plank. The only ones that will not require it are the few on the flat of the side of the boat just under the "sheer strake."

You will hardly be able to get these planks all out of one length. Not many boats nowadays are built that way, but where you do have to use two, make the seam where the two ends meet come midway between two frames and then rivet their ends to an oak block fitted snugly between the frames about half an inch wider on each side than the planking, so that the plank above and below will lap half an inch over this butt block. Common sense alone will tell any man not to make all these butts in his planking come in a line in one spot between the same two frames, but to shift the butts as far apart as possible, using the long length of a plank forward in one case and aft in the next, so that at least two planks come between butts made in the same frame space.

There are few places in this boat where the round of the side is so pronounced as to require hollowing and rounding the inside and outside of a plank so as to make it fit against the frame. Aft, on the quick round on the counter, and in the few planks that end in the hollow of the after frames this may be necessary. Never rivet a plank fast to the frame until its inner edge makes a perfect joint on the face of the frame. I know what you will be tempted to do. I have seen it done time and again, but those who did it always regretted doing so. That is to chisel off the face of the frame into a series of flats so that a flat plank will fit where it should be rounded. The result is the boat shows a series of ridges or if enough is planed off to make the plank show a smooth rounded surface the plank will be reduced to only about ⅝ or ½ inch in thickness, and as this is just where the fastening goes it is where the plank should have its full strength.

To fasten the planking to the frames use 2½-inch copper nails rivetted over copper burrs, and to make a good job first bore a hole so the heads of the nails will sink in or be counter sunk about half an inch. These holes, after the nails are rivetted up, are to be filled with cedar or white pine plugs dipped in white lead and tapped in over the nail heads, so that when the planking is finally smoothed off all will show a clean wooden surface and she will not look like a spotted pig, as she will if the nail heads are left flush. This is only done in very light rowboats or racing boats where the thickness of planking will not permit of countersinking; there the nail heads are smoothed off with a file, but *Mollyhawk* is not a racing shell.

Along the garboard seams, in the ends of the planks and such places as under a clamp where it is impossible to rivet up a copper nail use 2¼-inch galvanized iron boat nails, but bore for them just the same. Do not try to clout them in with a hammer for if you do you may spoil a plank that has taken you considerable time to shape, due to the nail buckling over in the hard oak and splitting the plank.

When all the planking is on, calk each seam carefully with boat cotton spun out and rolled to suit the size of the seam and paint each seam with a thin white lead paint.

This will stick the cotton in and hold it while you proceed with the rest of the work and makes the putty stick when you come to putty and paint the outside.

Before laying the deck, while the hull is all open, is a good time to put in your engine bed, line up and connect your shaft and install the motor. Make a templet for your gasolene tank and until that arrives fit in all the necessary floor beams both for cockpit and cabin floors. All can be fastened down with the exception of the forward cockpit floor beams. These will have to be left until the tank is in, so build a platform and sides to securely brace this tank in its proper place.

The floor beams for cockpit floor and cabin floor need not be dressed stock; that is, they can be left unplaned as they are entirely out of sight, the cabin floor, of course, being laid absolutely level; but the forward and after cockpit floors which are exposed to the weather would be better off for a slight crown so the water will drain to either side where lead pipe scuppers are to be fitted down and out through the outer planking so the rainwater will run overboard.

The main cabin floor can be made of wide stock, that is, boards 6 or 8 inches wide nailed down with the exception of a loose trap down the center. These will permit getting into the bilge of the boat if occasion should require your cleaning out the limbers, or to clean out the bottom when you lay her up. The two cockpit floors should be only 2 to 4 inch strips, the narrower they are the more yachty is the appearance. These decks should be caulked, payed and puttied and a rabbeted oak sill set in white lead and nailed down forming a sort of frame to receive the lower end of the cockpit staving. It takes a little extra work to get this rabbeted sill out and most amateurs, instead of doing so, will be tempted simply to nail a cleat on deck and then nail staving against this cleat; but the latter is very apt to leak, while the former insures an absolutely watertight job, and if you have ever lain in a bunk and felt the cold drip from a leaky deck you will know what this means. Take time and do it right now. You cannot change it later without a great deal of trouble and expense.

The laying of the side decks is slightly different. An oak edge-piece (sailors call it the "covering board") about 4 inches wide, the same thickness as the deck, which should be about one inch thick, is to be fitted so its outer edge is even with the outside of the planking. From there in the deck is laid in narrow strips. The forward and after ends of this decking are nailed to oak cleats fastened to the side of the after end of the house and the forward end of the after deck.

From the end of the raised deck the main cabin is made of two built-up sides consisting of a top and bottom rail with vertical stiles mortised and tennoned into them, forming the windows as shown in the plans. These sides are fastened to the deck by rods of five-sixteenths iron going through the lower rail through the deck and the end of the oak deck beams. By making these sides of 1-inch stock you will have wood enough to dovetail the ends of the short cabin beams into them. Here, also, it is customary to run one or two beams clear across from side to side to hold the sides accurately in position until all the others are in place, the deck laid and the sides of the skylight erected. Then, when the boat is secured by the skylight beams going across, these can be sawed out and their ends, as well as all the other beams, covered by a 3-inch by ⅜-inch finishing strip or, if you do not object to the ends of the beams showing, you can round off the ends of the beams with a chisel and let them show.

The construction of the cabin skylight is just the same as the cabin sides, although, of course, it is longer. The beams are dovetailed into the sides just as the cabin-house beams were, the deck laid in strips of white pine about 4 inches wide by ⅝-inch thick and the whole covered with canvas, just as the main deck was when it was laid, the canvas being held at the edges by a row of copper tacks and the ragged edge of the canvas covered by a half-round oak molding.

You have probably seen boats whose cabin-houses were defaced by dirty black stains running down from under this molding. If you will shellac the inside of the half-round molding before you put it on you will not have this difficulty.

The after end of the cabin, the forward and after cockpits, are to be built of ¾-inch tongued and grooved cypress staving about 2½ inches wide bradded to the sill-pieces and edge of the deck. Do not drive your brads straight in. Put them in on a slant and they will pull the staving and hold it much tighter. Punch the nail-heads in and finish the holes with a bit of putty.

How to Build "Mollyhawk"

Part III

By C. G. Davis

ANYONE capable of building *Mollyhawk's* hull knows enough to lay a quarter-inch tongue and groove cabin floor, and to build the seats and bulkheads either of one wide board set on edge or by first building a spruce frame and staving up the front with cypress or yellow pine tongue and groove staving in narrow widths. Boards two to three inches wide with their edges champhered so that when they come together they make a "V" shape seam is just as good and far cheaper to build than a transom front all formed of panels or other forms of expensive joiner work. All such ginger bread work, while it makes a boat look a little more stylish is no better than a plain pine board painted or a tongue and groove staving as we have suggested, and you can refinish the latter with one half the trouble and expense of a fancy paneled transom. The advantage of the "V" seam formed by the champhered edge is that it hides any slight unevenesses in the thickness of the seam where the two boards come together, which is not the case where the boards are square edged and, especially, if it is painted white. But as to just how you finish the interior of your boat, whether you use expensive joiner work or the plain cheap kind does not in any way affect the serviceability of *Mollyhawk*. It is a matter of personal taste and entirely up to you to say just how much money you care to spend on it. You can use cypress at about 5 cents a foot or use bird's eye maple or Circassian walnut at about 30 cents a foot. The same thing applies to the hardware below decks. Some owners will use the ordinary lacquered iron door-knobs and locks and drawer-pulls. Others will use glass ones, and others solid brass ones, the latter are far preferable, but when you buy them make sure that you are getting solid brass and not iron simply dipped in brass, as much of the boat hardware now sold is made. You will find, in a couple of years, the rust will strike through and your boat hardware will be anything but a thing of beauty. Especially is this true of the deck fittings, such as flagpole sockets, chocks and other deck plates, although my choice for such fittings would be galvanized iron in preference to brass. They are just as strong, if anything, stronger than brass, will not look so shoddy as brass does when it becomes tarnished, and if for any reason the galvanizing does get nicked you can retouch it with a bit of aluminum paint and make it look as if it had just come from the store. Let me call your attention to one apparently insignificant point about fastening on your deck plates and that is to see that the screw heads fit perfectly into the countersink bored in the chock or deck plate, whatever the fitting may be. Sometimes a hole for the screw is bored vertical, while the face of the deck plate is slightly beveling. The result is that one sharp corner of this screw-head sticks out like a knife, and will cut you when you polish the brass, and on which strings of cotton waste generally cling, looking anything but pretty. Use a metal countersink in your brace and bitt and ream out the holes until the screw heads just fit flush with the surface of the metal. Do not leave the heads standing up, and, on the other hand, do not let them sink an eighth of an inch or so below the surface of the metal. In that case you should use a screw with a larger head to fill up the hole, for such holes form puddles for the dirty brass polish or water.

There is one little point in the construction of *Mollyhawk* that I wish to draw your attention to particularly, and that is the knee on the after quarter, just above the half round moulding. This is called a quarter badge, and it is just such little fittings as this that set off your boat and add to her shippy appearance. Do not try to make this out of half-inch wood and plaster it on, for it will not stand. It is not like putting interior trim in a house, but get it out of one thick oak or hackmatack knee, as shown in the accompanying sketch, in which you will see that the knee itself is about two inches larger than what shows on the outside. By making the knee about two and a half inches thick you can cut a rabbitt in it and fasten the ends of the planking to it, leaving the little quarter badge extending out about a quarter of an inch beyond the planking and yet it will be solid enough not to curl or crack in the weather. While we are talking about knees just consider the two little sketches here shown. Most people do not think the mere outline of a knee has anything to do with a boat's looks, but in this they are wrong. Just as quickly as a house architect would notice a house built without eaves, so can a man used to water and ships spot a clumsy, amateurish shaped knee as shown in the upper figure. Such a knee, while it might be useful and appropriate in building a chair or a table, will make *Mollyhawk* look clumsy if it is used at the after end of her cabin or the forward end of the little raised deck aft, the turtle deck, as you might call it, where on either side a knee is shown which fills up what would otherwise be a very awkward looking square corner. Make the lower arm of the knee longer than the upright end and of some such curve as I have here shown.

Another little detail to which I want to call your at-

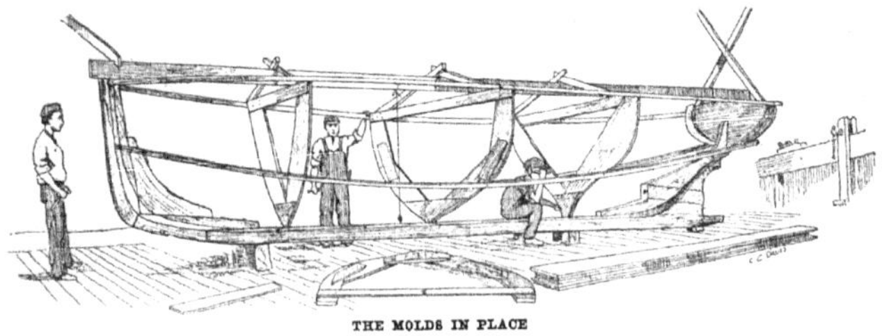

THE MOLDS IN PLACE

tention, and which applies not only to *Mollyhawk* but to all boats, is the quarter bitts aft and heavy mooring bitt forward. This forward one should go down and be mortised into the forward deadwood, although many people only use a short bitt and key it fast on the under side to an oak block which fits from deckbeam to deckbeam. This, to my way of thinking, brings too much strain on a yacht's deck for the main mooring bitt forward. Such a

PUTTING ON THE RIBBANDS

style of construction is all right for the small quarter bitts aft, but I would not advise its use forward. How many people have ever considered the reason why the edges of the bitts were champhered off the way they are on ships? Very few, I'll guarantee. But when I explain the reasons for it by means of the diagrams, A. B. C. D. and E. a blind man can see the point. I have seen bitts rounded off into all manner of fancy shapes, the man who did it evidently thinking that the idea of champhering a bitt was to make a fancy piece of furniture out of it. The real reason is this: The head of a bitt, as shown in figure A, is there to make rope fast to, and, naturally, the strongest part of this bitt is right at the deck. The higher up you go the more leverage the anchor cable has to break it. For that reason the champher is cut at such an angle as will make the cable ride down, and ride is the nautical word for slide, close to the deck. If you ever go to sea on an old sailing ship, where nearly everyone of the many ropes has to be coiled down over belaying pins, you would soon notice that on a belaying pin, shaped like Figure B, you can lay fake after fake over such a pin and they will pile up clear to the top without sliding off, as shown in Figure B. While with one shaped like C, when you get near the top the upper fakes of rope will begin to slide up over the top, as in Figure E, and you cannot coil nearly as much rope over such a pin. I have seen an old, deep water mate go along a ship's bulwarks, and every pin he found patterned after that shown in C he'd heave away to leeward with a deep sea blessing on the head of the man who made it.

A precaution to be taken when you are building the deck frame of your boat is to fit inch and a half oak blocks, snug between the deck beams and nailed to the same so they will come underneath the deck wherever there is to be a deck fitting fastened above. Do not trust to the deck. Soft white pine will never hold the screws, nor is it a good practice to put a cleat on deck so that only one screw comes into a deck beam, and you trust to that one good fastening to hold.

The finishing off of the boat after she is all planked and decked is most important if you want a good looking boat. The trouble with most amateurs is that by the time they get this far they are so anxious to get their boat afloat that they do not take the time to properly plane off, sandpaper and otherwise prepare the wood to properly receive the paint. Do not shirk this part of the work. Keep at the planing off of the seams and planking until all humps and hollows have disappeared and the plank, anyway you bend a small batten around its surface, shows absolutely fair. When it has been planed as true as is possible, start in with coarse sandpaper, folded over a block of wood and scrub the plank crossway to the grain until every plane mark is obliterated. Then, with finer sandpaper, rub it fore and aft, cutting out the marks of the heavier paper. Then, and not until then, is your boat ready for paint. In a boat where wooden plugs have been fitted over her planking you can go still farther by taking a bucket of hot water and a big sponge and sponging over the entire planking from deck to keel on both sides. Your boat has to be wet sometime, and the wood and the plugs have to swell. This sponging process makes the wood go through it's swelling before she gets overboard and shows particularly in the case of the plugs, which, owing to the wood having been slightly compressed when driven in with a hammer, is more apt to expand than the planking, and you can go around your boat as soon as the wood has dried with a chisel and shave off dozens of plugs that have swelled out a sixteenth of an inch or so beyond the surface of the wood. You can imagine what this would have done had you first painted your boat. For this reason many experienced boatmen never attempt to finish up a brand new boat as soon as she is built. They launch her and use her a month or so, then haul her out and allow her to thoroughly dry and then put her through the finishing process of sandpaper and two or three good coats of paint, for, by that time, the wood has come and gone all it will, due to swelling and any little straining the boat may do until she gets swelled up tight and solid has been done. Now, when she is finished, she will last for years, only requiring the surface of the paint to be replaced where it wears out.

For those who do not understand just how the line-up to which the copper is to be painted, for in *Mollyhawk* we

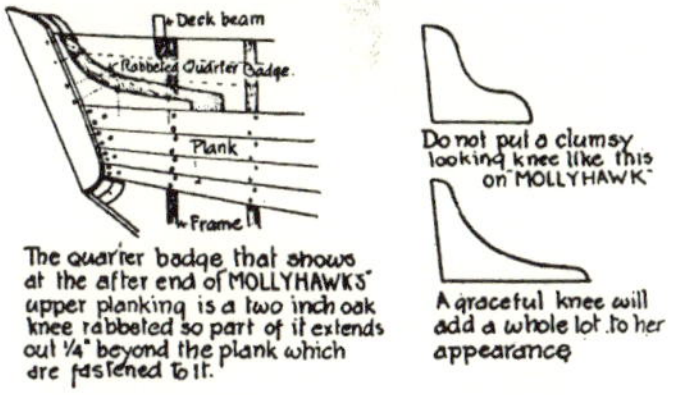

show what is called a boot-top, that is, several inches of the copper paint shows above water when she is afloat, a few words on this subject may be of assistance. With the boat set absolutely plumb, tack a straight edged board across the bows at the height you want the boot-top forward, and another across the stern the height the boot-top is to be raised there. Between these two, just so it clears the side of the hull amidships, allow a fish cord to sag until it gives you the proper height amidships, which, as you will notice, is lower than at either end. Then, with a long spirit level or a batten of wood and a short one, you can go along this line at intervals of every foot or so and mark spots on the planking to correspond with the height of this line. Then tack a batten, carefully sighting along it as you do so, to see that there are no unfair kinks in it, and with a race knife or the point of a brad awl cut or scratch a light groove along in the planking. Many a man before you has made the mistake of simply marking this with a lead pencil, which the first coat of paint has obliterated, for even a scratch, in time, becomes lost to sight through being filled up with the paint. It is a good practice to always keep this mark visible by re-scratching it occasionally, for nothing looks worse on a boat, as you yourself may have noticed, than a crooked wave-like line, where the two paints meet.

While it does not matter, if the boat is to be used in fresh water, whether the bottom be painted with a copper compound or not, it does make a great deal of difference if she is to be used in salt water where the torredo works such havoc in boat's planking by eating innumerable holes in it. Copper paint is the only thing that will keep this destructive little worm away and for that reason a great many people believe that copper paint should be put right onto the bare wood so that the copper can soak into

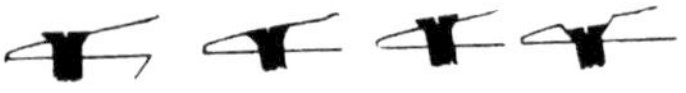

Hole for screw not bored at the right angle — Screw put in right — Screw head not countersunk deep enough — Screw head countersunk too deep

the pores of the wood, but as it is the liquid that really goes into the wood, depositing the copper on the outside of the planking, it is very doubtful whether this method has any virtue in it or not, or whether the copper be applied on top of a fine coat of lead. One thing we do know, and that is that the bottom should be kept completely covered with some copper paint and not allowed to chafe to the bare wood.

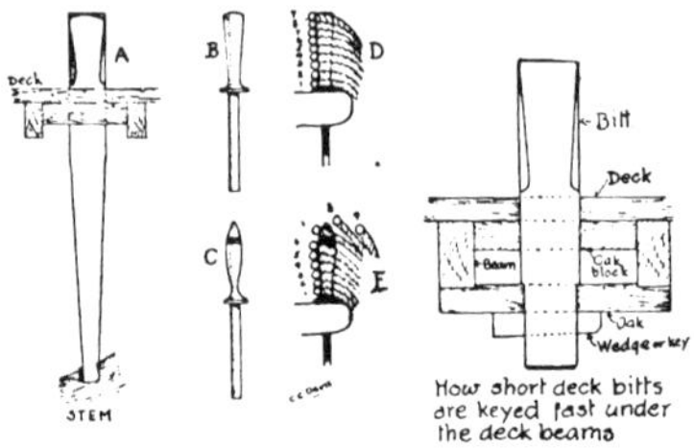

How short deck bitts are keyed fast under the deck beams

The finish of *Mollyhawk* I am going to leave entirely to those who build her. I do not know of any business that has so many conflicting opinions as that of painting a boat. Of course, I have my own views on the subject, but I can take you to another yachtsman who has had equal experience and he may advocate an entirely different manner of painting the boat. Some want a white painted top side. Others stoutly condemn it and say any color but white should be used. Some want varnished decks. Some would not have a varnished deck. Some will swear by one brand of varnish and some by another, all the result of personal experience on their part and more than likely the different opinions have been the result of accident more than anything else.

So, paint *Mollyhawk* any color you like; you'll do it to suit yourself, anyhow—you've a right to; she's yours.

How to Build the "Beaver"

A 23-Foot Cabin Cruiser. A Big, Roomy, Safe and Seaworthy Little Craft. Just What the Boys Want

BY C. G. DAVIS

The "Beaver" As She Will Appear Afloat

TELL you how to build a 23-foot cabin cruiser? Surely I will; that's the easiest part of the business, but it is up to you amateurs to do the real hard part of the work, the cutting out, fitting and fastening. And yet while it is hard work, it is one of the most enjoyable kinds of labor. You see the ship which is to be your future home grow by your own toil and labor. There's some satisfaction in building a boat that one is to use for his own, which is lacking when one builds as a business and never expects to see how the boat, over which he has toiled for weeks, behaves herself when in a sea.

To those men who have years of experience on the water, the *Beaver* will appeal with all her good points. Those who are newcomers to the pleasures of motorboating and who make their criticisms with no actual experience to back them, may say she is too wide or too high, or too something else, but these men I ask: "Have you ever been aboard of a converted Cape Cod catboat?" If not, you have no right to criticize this plan. Don't imagine that, because you have had a sail on a narrow motorboat and enjoyed it, the enjoyment could not be magnified by a trip on such a boat as *Beaver*. She has every requisite of a comfortable little cruiser for two or even four men or boys to go off for weeks at a time and live in comfort. Her beam gives room to move a step or two sidewise, and one is not confined to a narrow foot well such as ordinary small craft have between their transoms. Some may say her beam will make her a poor sea boat, but any day *Beaver* cannot go out no boat of her length will care to go.

Every man imagines his own boat is laid out better below decks than any other boat he ever saw, but let him look over *Beaver* and see if he can get any more room than has been put into this 23-foot boat Every inch of her has been utilized, from the coat room up in the bows to the ice-box and lazarette under the after deck. Two comfortable seven-foot transoms, two feet wide, afford permanent beds in the cabin, while out in the cockpit two more may be made up with curtains buttoned down to the sides of the awning, and berths made up on the long cockpit seats. A toilet room fitted with every convenience is shown, but this, of course, an owner can dispense with if he does not care for it. We have shown it to prove that it is possible to get such a room in the boat. Also on the starboard side a completely equipped galley is fitted in at this part of the boat, where the heat and smoke from it can best escape out the companionway, and not heat up the cabin. The engine can be gotten at, and yet is completely out of the way, just its flywheel protruding a few inches into the cabin to enable one to start the engine. Two cylindrical tanks, one for water and one for gasolene, are fitted in chocks and securely lashed under the cockpit sides, where they are least apt to affect the trim of the boat, as their weight decreases, and at the same time are spread apart to assist the boat in swinging with a slow, easy recovery in a sea way

What the *Beaver* will look like when completed is shown in the sketch and accompanying plan showing the deck view and appearance above the waterline. The resemblance to a roomy Cape Cod catboat is noticeable, and is intentional, as that type, when converted into a motorboat, has proven itself a most satisfactory cruiser; but with the similarity in bottom the parallel ends: *Beaver* is a typical motorboat from there up, yet one that has elbow room seldom found even in motorboats much larger than 23 feet. It is queer that with the hundreds of examples before us of converted sailboats' hulls, and the speed, weatherliness and comfort thus attained, no one has attempted to design boa. along these lines.

But let's get busy, for the plans speak for themselves, and any one who wants to build does not need convincing that she will be a good boat; too many of MOTOR BOAT'S subscribers have been waiting for such a craft, as the many letters received testify

If you have a shop, shed or barn where you can build your boat under cover, so much the better; if not, and you have to build her out in the weather, be sure and shovel off the loose top soil and get your keel blocks firmly planted on hard ground, so they will not settle under the boat as she accumulates weight.

The keel is a straight piece of oak 22 feet 11 inches finished length, 3 inches thick by 4 inches deep. Pick out a straight, clear-grained piece, as free from large knots as possible. If you order this stock for your keel at a mill, order it dressed, as it is termed. It will only take the mill a few moments to run it through a power planer, and then you will have a good, smooth stick. If not, you will have to scrub it off with a plane, and this is

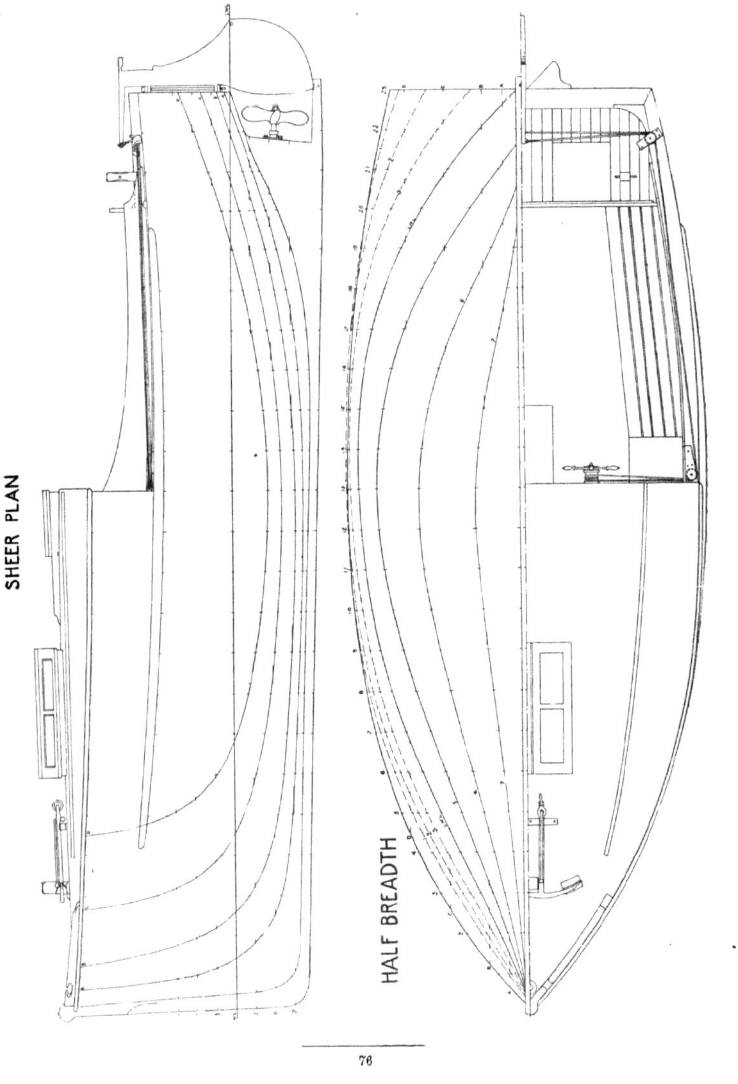
SHEER PLAN
HALF BREADTH

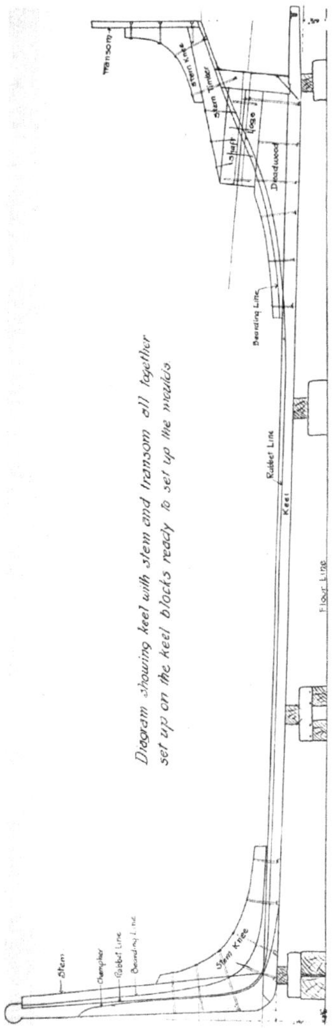

Diagram showing keel with stem and transom all together set up on the keel blocks ready to set up the moulds

considerable work. Then cut it to the shape as shown in the drawing, which as you will notice is very little work indeed, and we have purposely made it so. In fact, everything about the boat has purposely been made simple and easy for amateurs to build. The rabbet, for instance, instead of being a dug-out rabbet, as it would be with a log keel, in this boat amounts merely to a chamfering of the top edge. The angle at which this chamfer is to be cut is explained in the accompanying sketches, Figs. 1 and 2, on page 14, and is found by taking a piece of wood the same thickness as your planking and butting it against the top corner of the keel at the angle the molds make where they fit on the keel.

First cut notches opposite each mold; then with a batten draw connecting lines and cut the rabbet from one spot to another, continuously, but do not try to finish it where it runs into the stem and where the deadwoods lap onto the stern. Leave the rabbet uncut there until you have the deadwoods bolted on. Then, with a batten, draw the line in far across the stem and cut the rabbet. It would be well in all cases to paint the various parts of the keel as you get them out with a thin coat of lead paint, to prevent their checking as they dry out, which they are bound to do unless your wood is already perfectly seasoned.

Now, get out the stem. This also is of oak, and requires a piece of wood 6 feet 6 inches long, 3 inches thick and 10 inches wide at its widest part, tapered to 6 by 3 at the top. Smooth off the stem the same as you did your keel on both sides, but remember one point, and that is always to work all the deadwoods, stem, etc., from one side only; that is, if you are squaring up from the starboard side of the stem, keel and deadwoods, in squaring the edges, always apply your square to this face, making everything square to it. If you try your square first on one side and then on the other, you may find unevennesses of the pieces of wood that will throw the square off

Get the lower end of the stem so that it fits, notching

BODY PLAN

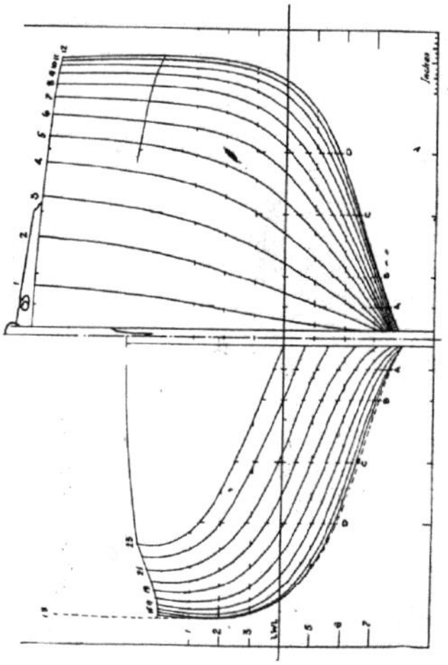

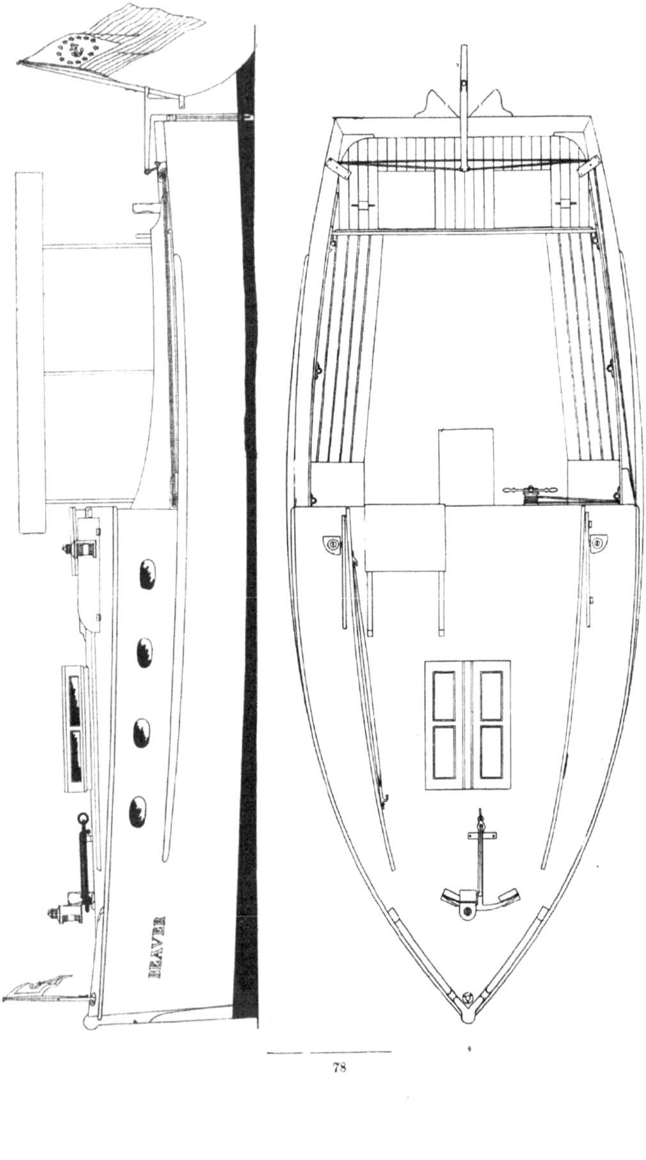
BEAVER

over the end of the keel as shown in the plan, and then lay out the lines marked rabbet line and bearding line. Between these two you are to chisel out the notch termed the rabbet, into which the ends of all the planking are to fit. You will notice that these two lines are narrowed together at the top and spread apart considerably at the bottom. The reason for this is the planking approaches the stem at a much sharper angle near the bottom, the same thickness of plank requires a broader surface fore and aft. The planking in this boat is three-quarters of an inch thick and this rabbet and bearding line has been figured on our plans for this thickness of wood, and you will find, if you take a small piece of ¾-inch wood as a templet, when you chisel out the rabbet and apply it at the different bevels at which the different waterlines approach the stem, that you will require just the widths shown on the stem for the rabbet. Figs. 3, 4 and 5 illustrate what will happen if you do not get this rabbet cut just right. In Fig. 3 the plank is shown as it properly fits into the rabbet. The outside of the plank finishes flush with A, which is the rabbet line, the inside finishing flush at the spot B, representing the bearding line. If you do not hold the little templet of the planking at the right angle, Figs. 4 and 5 show what will happen. In Fig. 4 the templet has not been applied at enough of an angle, and in Fig. 5 the angle is too acute, and you will notice the edge of the plank does not fit in flush with the surface of the wood. This same principle applies aft, of course, as well as forward, where you get the rabbet line across the deadwood shaft log and stern timber. Be sure that the after side of the stem is cut perfectly square with the working face, and then get out a stem knee, as shown in the drawing, and see also that the edges of this are perfectly square. If they are not, when you come to rivet the stem and keel together to this knee, they will not be in a perfect line, but will stand decidedly crooked and twisted.

Owing to the different lengths of bolts needed in the keel and deadwoods, carriage bolts are very seldom used. When you come to bolt the clamps and frames together, you can very often make use of carriage bolts, but here you need galvanized rod iron. Any firm that sells boat hardware carries this in stock in all sizes from ¼-inch up to 1 inch or more in diameter. For this boat you will require two rods of about ⅜-inch iron and about two dozen clinch rings for the same. Bore your holes as shown in the plan where the bolts are to go, measure the exact length with a little thin piece of wood, and get the bolts just the right length to reach from one edge of wood to the other, allowing the heads on the outside to countersink to a depth of about ¾ to 1 inch, and this countersink should be bored with about ¾-inch bitt. These holes, we might here say, are afterward to be filled with wooden plugs which you can also buy at a boat hardware store. Hold the bolts in a vise and rivet up one end so you form quite a good-sized head on it, and while doing this you will notice the quality of the iron you are using, and it should be such that as you form the head, it will not flake off and fly in pieces, but will stand turning over and forming a good, strong ridge.

Paint the adjoining faces of the stem and knee with a good, thick mixture of white lead paint just before you bolt them together. Then drive the bolts in from the outside, put a clinch ring over the inner end, and rivet them up firmly, having some one hold on against the outer end with a good heavy weight, such as the pene end of a top maul. In regular boat shops they have a five or ten-pound block of iron with a hole drilled in it, into which a stub end of an iron bolt is put, which they hold against the bolts, but this, of course, amateurs would hardly have at hand. There is one bolt, the forward one in the keel, that does not go all the way through. This is called a driftbolt. Bore for this the same as you did the other, but be careful the bitt you use does not bore a hole so large that the bolt will slip in too loosely. If anything, it should be 1/16-inch smaller than the drift bolt, so that, as you drive it in it will hold and draw the wood tightly together.

The after end of the backbone of this boat is composed of several pieces of wood, and,

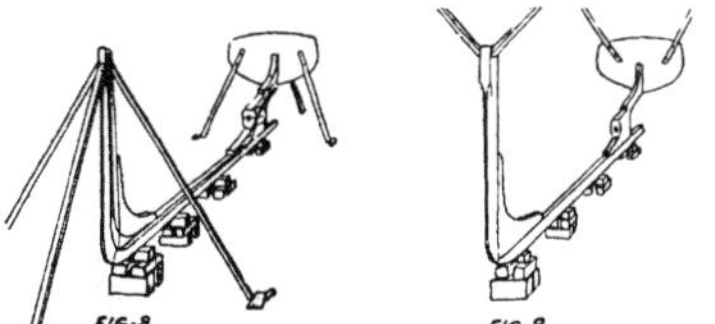
FIG-8 FIG-9

Figure 8 shows the keel braced to the floor, Figure 9 shows how much easier it is to work around a boat when the braces can be put up to the beams overhead.

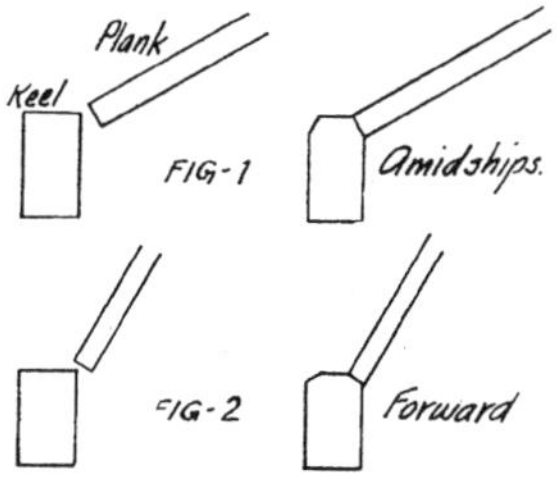

The rabbet along on the keel is only a bevel cut so the ¾ inch plank will fit square at the different angles in which it meets the keel. Forward the plank stand more vertical - amidships more horizontal.

FIG-3.

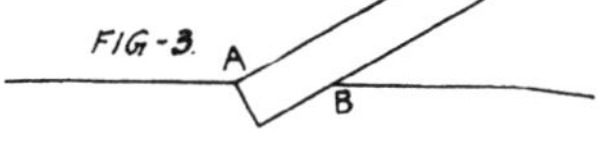

FIG-4

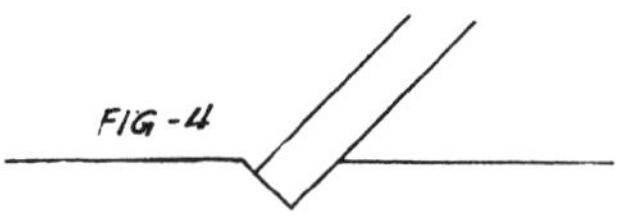

FIG-5

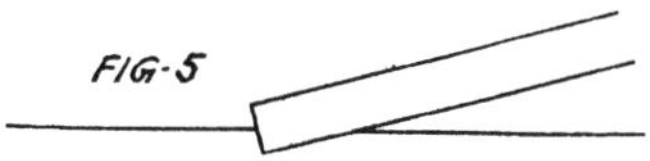

LIST of lumber needed to construct BEAVER										
ITEM	NO. OF PIECES	LENGTH FT.	LENGTH IN.	THICK. FT.	THICK. IN.	WIDTH FT.	WIDTH IN.	WOOD	REMARKS	REMARKS
Keel	one	24			3		4	oak		dressed
Stem	one	6	6		3		10	oak		"
Deadwood	one	10			3		9	oak		"
Shaft Log	one	5	6		4		5	oak		"
Stem Knee	one				3			hackmatac	3' arms	square
Stern Knee	one				3			hackmatac	2' arms	square
Quarter Knees	two				2			hackmatac	2' arms	Out square
Breast Hook	one				2			hackmatac	2' arms	In square
Moulds	one	12			1½	1		spruce		un planed
"	four	9			1½	1		spruce		"
"	two	14			1½	1		spruce		"
Transom	one	20			1½	1		oak		dressed
Floors	four	20			1½		6	oak		"
Deck Beams	ten	10			1¼	1		oak		"
Cockpit Beams	nine	8			1¼		2	spruce		"
Engine Bed	one	6			2		10	oak		"
Frames	six	10			1½	1	'	oak		dressed
"	six	8			1½	1		oak		"
Planking					¾			cedar	400 sq. ft.	dressed
Cabin deck					¾			cedar	120 sq. ft.	"
Covering board	three	10			1		8	oak		"
After deck					1		3	pine	70 ft	"
Cockpit floor	twelve	10			⅞		8	"		
Cabin floor	six	10			⅞		8	"	Tongue &	grooved "
Bulkheads					⅞		3	"	to cover	100 sq. ft "
Wearing strip	two	16			2		3	elm		
Moulding	two	14			¾		1½	oak	half round	
Clamps	four	24			2		4	yel. pine	or spruce	dressed
Clamps	Two	14			2		4	" "	" "	"
Mooring Bitt	one	6	6		4		6	oak		"
Quarter Bitts	Two	3			3		3	oak		"
Bow chocks	one	7			2		4	oak		"
Rudder	one	8			2	1		oak		"
Coaming	three	9			¾	1		qtd. oak		"
Companionway Posts	Two	3	6		2		4	oak		"
Seat frames		40			1½		2	spruce		
Seat slats	Ten	6			1		3	qtd. oak		"
Bunks - fronts	four	8			⅞		8	pine		"
" tops	four	8			⅞	1		"		"
" face	Two	8			½		3	qtd oak		"
Skylight ridge	one	3	6		3		3	pine		"
" sides		8			1½		8	"		"
" ends		5			1½		8	"		"
" frames		32			1½		2½	"		"
Companionway	Two	3	6		2		4	"		"
"		4			1			"		"
Deck battens	Two	9			1		1½	oak		"

owing to difficulties many amateurs have in boring their shaft holes, we have shown that style of construction in which the shaft log is built up of two pieces of wood, with one-half the shaft hole gouged out of each piece, and the two held together by iron dowels. Get these two pieces out first, and the piece of deadwood that is to go under them, as per plan, and then bolt them together, riveting up the bolts on the inside of the shaft hole, but be sure they are countersunk well below the shafthole, so as not to interfere with the sleeve when you come to put it through the shaft hole. Then bolt the stern timber to the upper piece of the shaft log, and the up and down sternpost which binds the ends of these deadwood pieces together. Be very careful that the holes for your dowels which extend in a row either side of the shaft hole in the shaft logs are perfectly in line, so when you set one piece of wood on the other, they will drive tightly together. The two shaft logs in this boat, owing to the deadwood being only 3 inches wide, are made of heavier stuff, 4 inches, beveled off outside of the rabbet line, so they face down with the 3-inch stuff.

When the deadwood is all bolted together, drift bolts and all, get out a stern knee, as shown in the plan. To this the transom is to be riveted. The transom is made of 1¼-inch oak, the shape shown in the plans, but, of course, you cannot buy a piece of oak as wide as is required for this. So you have to join two or more boards together to get the required width. This requires another nice piece of carpenter work to face up both edges so as to make a perfect seam, and then dowel the two together. Fig 6, on page 18, shows one method of doing this, that is, where they are doweled together, and Fig. 7 shows how an amateur who feels he cannot successfully make the dowel joint, can put them together by nailing cleats across the inner face of them.

It is generally customary to get this much of the boat all bolted together before the keel is "set up," as it is termed, and the manner of doing this varies according to the conveniences—or lack of them—at the builder's disposal. For instance, a man who has a shop with a good, level floor to work on, has many advantages over the man who has to work outdoors on the ground. All the former has to do is to lay a pile of blocking along the floor which will raise his keel 14 inches at the forward end and 6½ at the after end, as marked in the plan. But a man who has to do this work outdoors will have to scrape away the loose earth and get down to hard pan before he begins to build up his blocking. If he doesn't, he will find that as the weather comes and goes, rain and dry, the blocking will settle his boat all out of line.

The manner of holding the keel firmly to the blocking while you put up the molds and proceed to the building of the boat is also done in various ways. Fig. 8 shows the method where these braces are carried down to the floor or to the ground, as the case may be, while Fig. 9 shows a much better system of shoring the keel to the rafters or beams overhead, which leaves the floor perfectly clear for a man to walk and work around his boat without stumbling over the braces. The stem and transom should be braced both sidewise and fore and aft, as considerable strain is to be put upon both of them when you come to bend the stiff yellow pine ribbands around as you proceed with the work.

Where these braces are nailed fast to the upper part of the stem, called the stem head, it is customary to leave that part of the stem larger than is needed for the finished job, or to leave the stem head longer than is actually needed, and then when the planking is all on and you can dispense with the shores, the stem head, which at this part is full of nails, can be dressed down or if left longer, can be sawed right off and thrown away. In the case of the after end at the transom you cannot do this, so secure the braces to the inner face of the transom, where the holes will not show when you take the nails or screws, whichever you have used, out of the braces.

The temporary molds which are to serve as guides to give you the shape of the boat when framing, can be made of about 1-inch stuff if the boat is to be built with cold fitted timbers. By this I mean timbers bent over a

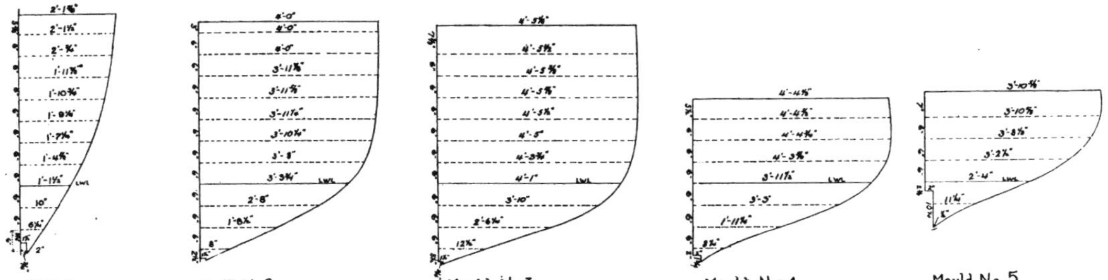

Mould No. 1 Mould No. 2. Mould No. 3. Mould No. 4 Mould No. 5.

Measurements for laying out the shape of the moulds used in the construction of the 23 foot Motor Boat "BEAVER."

Figures for laying out stem and deadwood

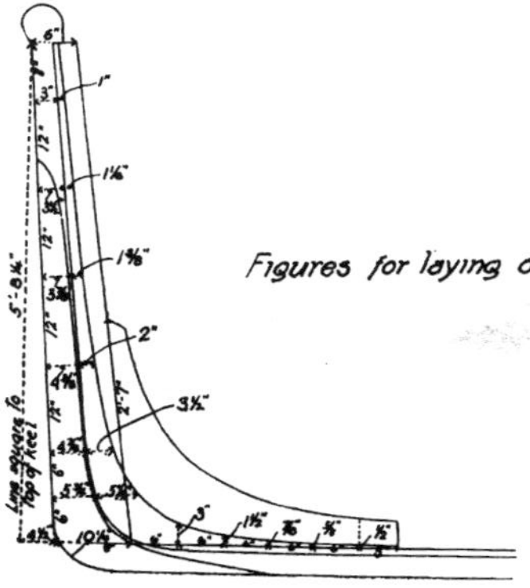

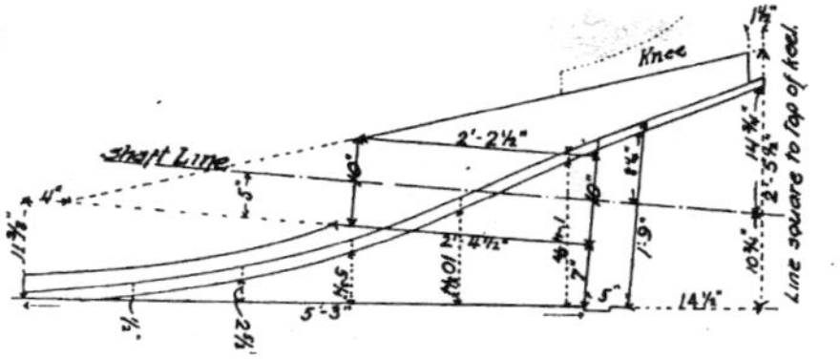

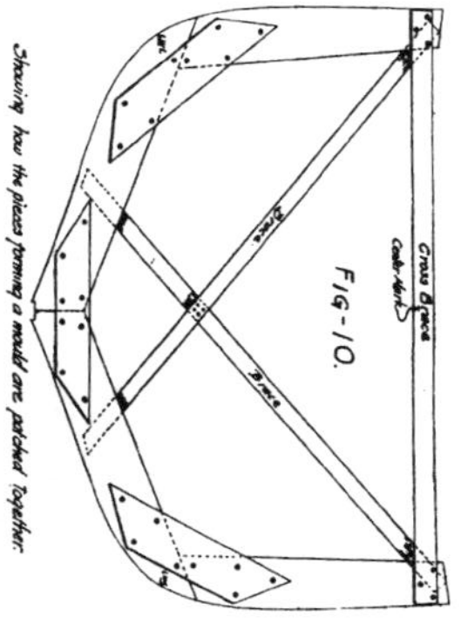

Showing how the pieces forming a mould are patched Together.

mold and when cold beveled and fitted in around the ribbands which run fore and aft along from one of these molds to the other, but if the boat is to be framed with hot timbers, that is, the timbers taken from the steam box and bent directly in around the ribbands, it would be advisable to make these molds of at least $1\frac{1}{2}$-inch or 2-inch spruce. It does not have to be a very good quality of wood, as it is all temporary work, and is of no use whatever when the boat is done.

A great fault with many of the amateurs in attempting to build a boat is in trying to do with too few of these molds. Some of the best boatbuilders in the country go to the trouble of making an individual mold for every frame, but this, of course, is not necessary in this case. I have shown five molds, Nos. 1, 2, 3, 4 and 5, and given all the dimensions necessary for determining their shape. Mold No. 5 is to be set up three feet forward of the after edge of the transom and the others each four feet apart. Molds 4 and 5 have their faces aft and the Molds 1, 2 and 3 are to be set with their faces forward. Mold 1 is set where frame 4 goes, No. 2 where frame 8, No. 3 where frame 12, No. 4 where frame 16, and No. 5 at frame 20. These frames will have to be left out until the molds are removed and then put in in their places.

Setting up a mould:- One man plumbs it while another nails the braces fast.

An amateur could save considerable time, perhaps, by cutting the boards out of which the molds are to be made so as to approximate roughly the shape desired, and measuring off from the center line the measurements given in the mold shapes, and drawing a batten through this, but while doing so, he should be very careful that the boards do not shift. If they do, the whole mold will be of imperfect shape and the boat will be unfair when you come to build her.

If you have a large, clean floor space, or the same could be provided by laying down sheets of heavy brown wrapping paper, it might be well to measure off these mold shapes on the floor. Then, by laying a row of nails so that their heads just came along on the line so drawn, the shape could be reproduced on to thin pine pattern boards by laying them carefully over the nails and stamping on them. This is the way it is done in a boat shop. These thin patterns are then cut with a draw-knife and again fitted back to the pencil marks on the floor. When they are found to be accurate, they are tacked together to give one side of the mold, for in boat work we only do one side at a time and reproduce the other side by simply reversing the pattern. The molds themselves can then be cut by laying this pattern onto the heavier stock out of which the molds are made.

In a professional boat shop, where a power-driven band-saw is at hand, the labor of getting out these molds does not amount to very much, but an amateur who has to saw them out by hand and drag them out with a draw-knife will think by the time he has got out his molds alone, he has done pretty nearly enough work to build the boat. If a mill is handy he might take his patterns there and have the molds sawed out.

When the two sides of the mold have been cut out, spread them out on the floor, one each side of a center line, as shown in Fig. 10, and cleat them securely together. The upper ends should be left about six inches longer than is actually needed in the boat. This is done so that the cross pall, as it is called, which in plain language means the brace across the tops of them, can be bolted fast to the molds, so the lower edge comes just flush with what is to be the sheer line in the boat. This you will find later will be a considerable help to you.

Another form of checking up the boat's molds is to mark where the load waterline and center line come. Then you can plumb your mold when you set it up on the keel by dropping a plumb bob from the center mark on the cross pall to make sure the mold stands plumb, and you can also, with a spirit level, see whether it is level across. You cannot be too particular in putting up your molds to see that all this work is very carefully done, if you do not want to build a lop-sided boat, and that is what you will surely get if you are not very careful in plumbing the molds. As all the strain of bending in the hot frames comes upon these molds, you can readily understand that they must be very securely braced, the same as the stem and transom were, the overhead method being far preferable. They can also be braced by running a fairly wide plank, say, a plank 6 or 8 inches wide, down through the middle of the boat, so one edge of this board just lines up with the center mark of each mold. This will help hold the molds in position, and at the same time is an excellent fore and aft brace for them.

There are several methods which can be used in framing the boat. Some builders use one method and some another. So, when I state in this particular case one method of how to build this boat, and you see a boatbuilder doing it another way, do not imagine that he doesn't know his business, or that I do not know mine. It is merely a matter of choosing which way we think will be the easiest under the circumstances. Builders, for instance, put up, say, four molds which show the shape of the boat at four points, bend very light spruce or yellow pine ribbands around these molds, and saw out the shape of each particular frame. Others bend a lot of frames over a mold that approximates very closely to the shape, allow these frames to set and become cold, and then fit them one by one, beveling them and straightening them out wherever necessary to give the shape, and riveting them up in pairs before they go into the hull, but when they do go in, they are perfect and complete.

If I were building a boat myself, this would be the method I should follow, but it is much more difficult, and for that reason I am going to tell an easier way in which the amateur can perform the same work, and that is, instead of bending light ribbands around your molds, you must get good, solid ones, say 2 inches square, of yellow pine, long enough to reach from one end of the boat to the other. If not in one piece, they can be reinforced where the two join together by a piece 6 or 7 feet long, to which they are riveted or screwed.

Bend about seven of these around the molds on each side, then heat up in the steam box as many timbers at a time as is possible, take them out of the steam box while still hot and bend them in at intervals of a foot inside of these ribbands, drawing them up to them by boatbuilders' screw clamps. This method does away with all that difficult beveling which the timbers that are fitted in cold are subject to, as here the timbers while hot can be twisted so that they fit flat against the ribbands. For this purpose, while you are bending them, provide yourself with a good-sized monkey wrench, which will give you leverage enough to twist most any bevel necessary in the most stubborn frame. The greatest difficulty in framing a boat this way is the liability of the frames to split or crack in two where they are required to take a quick bend, which in this boat I have attempted to dispense with, making the frames as easy a sweep as is possible, at the same time getting a well-shaped boat.

The hardest part of this frame will be found in the heads of the after timbers. There the curve is the quickest, and it may be necessary to construct a special mold which will give you about the curve required at that point, and bending the timbers over it to get that sweep. Then, if they will not fit in cold at the heels, re-steam them and bend them in where needed. The novice will find the subject of steaming frames gone into in detail in the MOTOR BOAT HANDBOOK, VOL. I.

In buying your oak for the frames, you do not want dry stock, but on the other hand it could be decidedly green. If it is too dry—and you will soon be able to tell if such is the case—the frames will crack right in two. You do not want any quarter-sawed oak; only the straightest grain boards should be used.

Do not attempt to use boards for frames that are full of knots. If you do, you will find there will be considerable waste, as the frames will break nine times out of ten where the knot distorts the grain of the wood, and you will only waste time and material in attempting it. If you bend in a batch of frames to-day, they will be set sufficiently by the morrow to enable you to take off the clamps, cut the heels so they butt against each other at the center of the keel, and then proceed to nail them securely to the keel with about 2-inch galvanized boat nails, and nail them lightly to each ribband so that you can do away with the screw clamps, and use them in bending in a fresh batch of timbers. It would be impracticable to have enough screw clamps to frame the boat at one time, and very few boatbuilders do so. In fact, by the time you have bent in six or eight pairs of frames, you could then nail them temporarily to the battens and remove the clamps to use on others.

Along in the way of the engine amidships, that is, between frames 12 and 16, bend an intermediate frame between each of the regular timbers, so as to reinforce the boat at this point, to withstand the vibration of her engine.

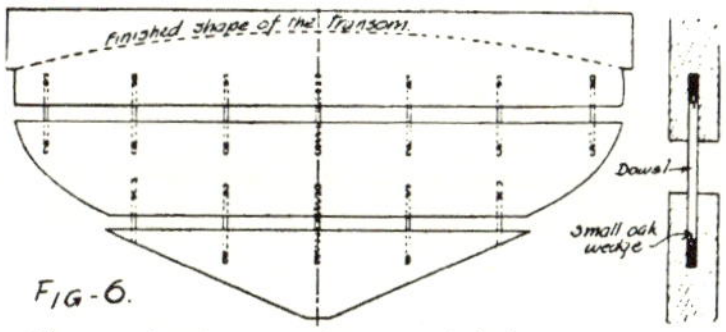

Showing how transom shaped out of three pieces of wood is doweled together and details of the dowel

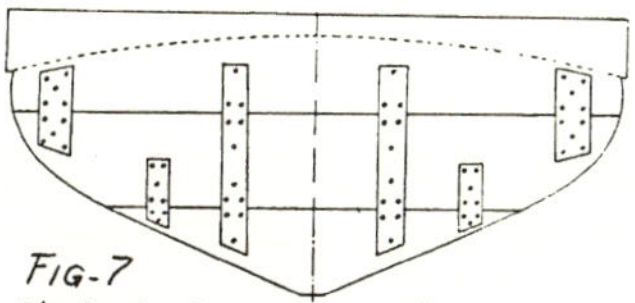

Showing how transom is sometimes held together by cleats screwed fast inside

If you wish to carry this doubling up of frames a little further fore and aft, it will not hurt the boat any, but at least put them as far as shown in the plan

When the frames are all bent, the next step is to reinforce them at the keel with sawed oak floors, 1½ inches thick. To save considerable fitting and to simplify the job, in this boat we have carried each floor up so that the top edges provide a level surface on which the cabin floor can be laid direct without going to the trouble of fitting in an additional set of floors

Some builders, instead of using sawed floors as we have shown here, take short pieces of the same stock that the frames are made of, say, pieces 4 to 5 feet in length, steam them and bend them in right on top of the frames, across the keel, connecting one frame with the other, and riveting them to the frames. If this is done, of course, additional beams have to be fitted to receive the cabin flooring, which our method of construction dispenses with. To get the shape of each floor as you are building it, stretch a chalk line fore and aft from stem to deadwood at the height shown in the plans, then take a thin—say ⅜ to ½-inch pine board—about 4 feet wide and 9 or 10 inches deep, lay it across the top of the keel, against the frame whose floor you wish to find the shape of, making sure that the top edge is level, and with a pencil mark along the outside of each frame on this board. You will find this a very quick and simple method of determining the shape, and with a draw-knife you can cut this thin wood pattern along the pencil line and use it to cut the shape of the floor out of the heavy 1½-inch oak. By beginning amidships, where the angle of the floor is flatter, you will find as you proceed forward from frame to frame, the one pattern can be used over and over again, a shaving being taken off each time as the angle sharpens up toward the bow, and a similar method will enable you to get the after floors.

Nail the heel of each frame to its floor with a 3-inch galvanized iron boat or wire nails. You may have to bore most of the way for these nails. If you do not, you may split the frames, which, after all the work you have gone to, is anything but pleasant.

It is a good practice to double up the ribband that goes along the head of the timbers—the sheer ribband as it is called—by bending one outside of the other and lag-screwing them together to the stem and stern and to each mold. Do not trust to merely nailing these ribbands to the molds, for, as sure as you do, you will meet with a catastrophe by having the ribbands spring off and get your frame all unfair. The last frame in the boat, frame No. 23—for our frames in this boat are spaced a foot apart—is to be screwed fast to the inner face of the transom, reinforcing the same so that when you come to put your planking on, you can put an alternate fastening, one in the frame and one in the transom, giving a double holding surface which at the ends of the boat is a very important part of the construction.

When all the floors have been gotten out and riveted to

the frames, you can knock out the temporary molds which were first erected on the keel, so as to enable you to put in the fore and aft yellow pine stringers. There are three of these on each side, one which goes along the upper edge of the forward timbers, forming a ledge on which the cabin roof beams land, another one running along at what is considered the sheer height, and another one, known as the bilge stringer, which runs along halfway from the keel to the deck edge. The purpose of these stringers is to stiffen the frame, and they should be of clear yellow pine about 2 inches thick and 4 inches wide. To make them a little easier to bend in the ends where the curve becomes quite sharp, it is customary to taper these stringers to, say, 1¼ or 1½ inches thick by about 3 inches in depth. If you do not taper them, put the forward ends in the steam box and make them soft and pliable. If you don't, you will never get them around the curve without breaking.

If you have been careful in bending in your timbers, and careful that the timbers were of the same size, all 1½ inches thick, before bending them in, you will find that these clamps will fit fair and true on the inner edge of each timber. If they do not do so, shave down the high timbers until they do make a perfect fit on each one. If you do not, the timbers will be pulled in and out, forming a very irregular and unfair side line. Here is the place where you can use carriage bolts to advantage, if you want to. Bore from the outside, countersink the heads of each one of the carriage bolts into the frame, and set the nut up tight on the inside of the clamp, cutting off any end that may protrude, and tap it a little to slightly rivet it just enough to keep the nut from untwisting and loosening. Do not put these bolts in a direct line. Put one, say, near the top of the clamp, the next near the bottom, etc., staggering them alternately up and down.

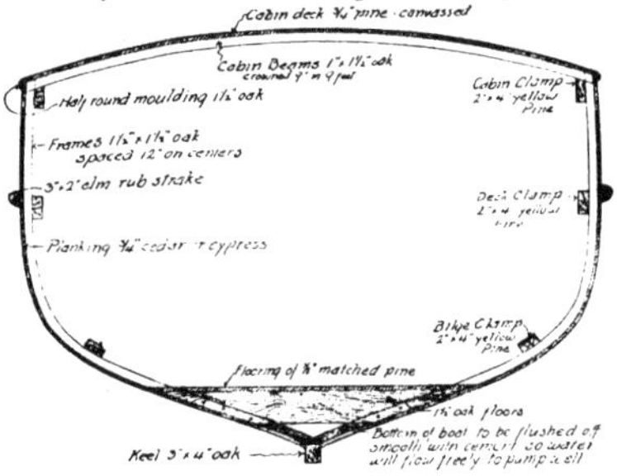

As the cabin beams are to be 1½ inches deep, be sure to set the upper clamp that distance below the edge of the boat, plus the thickness of the cabin top, so that when these beams are nailed to it, their upper surface will just come flush with the heads of the timbers, which, of course, are underneath the deck. Now, to stiffen the three corners which this boat has, that is, the stem and the two quarters aft, knees are fitted in. In the case of the forward one, which is technically termed a breast hook, many people do not go to the trouble and expense of putting in a regular natural growth knee, but, instead, merely fit in a three-cornered block of oak, the thickness of the deck beams, 1½ or 2 inches, which is notched around the head of the first timber, or you can cut this timber off flush with the bilge clamp, and bolt the breast hook to the stem by a diagonal fastening or with a long bolt bored clear through from the forward side of the stem to the after side of the breast knee.

Aft, where the angle is considerably more open, a knee is far preferable to a mere block of wood. Their arms are short. In our case it only requires a 22-inch knee of about 2 inches thickness. This knee is fitted to the inside of the transom and riveted fast to the same, spiked to the upper edge of the deck clamp, upon which it rests, and when the plank goes on, you can get a few fastenings of the top strake into this knee.

Part II

BE sure to brace the frames the same as you did the molds, before you remove them to put in frames in their place, because the boat will be subjected to considerable strain in being planked. For this same reason it is better to put in all the deck beams, cabin beams, and even the cockpit floor beams can be fitted, but not fastened, as the latter would be in your wav when riveting up the plank fastenings. This work can be done so much easier now than after the hull is all shut in with planking.

The cabin beams are of oak, 1 inch thick by 1½ to 2 inches deep, cut with a sweep that raises them 9 inches in the width of 9 feet. You can either sweep this curve with a long wire or a batten about 13 feet 3 inches long, or you can lay it out by taking the measurements off a smaller circle swept with a 9-inch radius, as shown in Figure 8. The quarter of the circle is divided into four parts, a, b, c, and the base line in four, d, e, f. Then, on a thin (½-inch) pine board 9 feet long, snap a chalk line for a base line, and in the middle measure up 9 inches, then from the middle toward each end divide the 4½ feet into four equal parts, d, e, f, and lay off the distances a, b, c, etc., on the bevels, as found in the small circle. Do the same reversed for the other side and then, by bending a batten through these spots, you get a true curve for a pattern or beam mold by which to mark out your cabin deck beams.

The beams across the after deck can be cut from the same pattern, but those across the cockpit floor should be laid out to a very much flatter curve, one with only about 2 inches round in the width of the cockpit.

Many amateurs, I know, will make them straight across, because it is easier to do so. The advantages of having them curved a little is that they will drain off any rain water quicker than one laid dead flat.

Now comes the job of planking up. This frightens many amateurs, but, as a fact, it is one of the most interesting parts of all, when done right and studied as you go on with it. Like an economical tailor, you can cut your planking so as to make it go a long ways or, if you don't use your head and good judgment in selecting the boards, you can waste a lot of cedar. There is, at best, considerable waste in planking a boat, about 25 per cent., generally, varying more or less as the shape of the boat approaches or departs from a round, barrel-like shape.

The principle of planking a boat is the same as the construction of a barrel. The barrel staves are all shaped wider in the middle and narrower at the ends, only in a boat the ends are not all the same size as in a barrel. After one or two planks have been fitted onto our hull, we will return to this barrel principle and you will realize the similarity in construction.

The first, and perhaps the hardest, plank of all to fit is the one next to the keel, called the garboard strake. The difficulty is principally in not knowing how, and in trying short cuts, but I can assure you that the ancient boatbuilders have in the past centuries found and made use of all short cuts, and so don't you try to find out any more. I know I did when I built my first boat, but I had to come back to the orthodox method of "spiling."

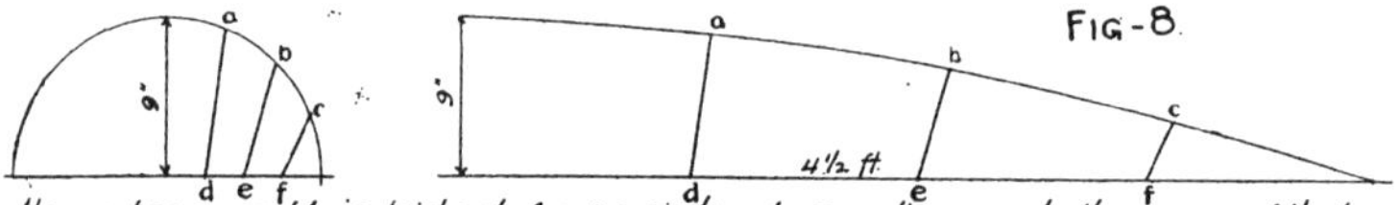

How a beam mould is laid out from a circle whose radius equals the crown of the beam.

Spiling for the shape of a plank consists of tacking a thin board, say, ⅜-inch thick, to the frames, so that its lower edge, in near the rabbet, along on the keel, is cut so that it roughly fits to the curve the rabbet takes onto the stem, as shown in Figure 9; with another such board tacked to the after end and the two lapped and nailed together amidships, or one long plank, if you have it. With a pair of dividers set to span the greatest distance between this "spiling staff," as it is called, and the rabbet, prick off a series of spots to give you the curve necessary to cut the garboard to, so it will fit. At the ends, where there is considerable curve, lay your rule across at intervals of 3 or 4 inches, and mark a series of lines, to give you the direction in which the measurements are to be taken off and measured back on. Take plenty of measurements around these curves, but along amidships, where the rabbet runs in a straight line, you will not need so many. One every foot or 18 inches will be enough.

Then carefully remove this thin pattern (the rule staff)

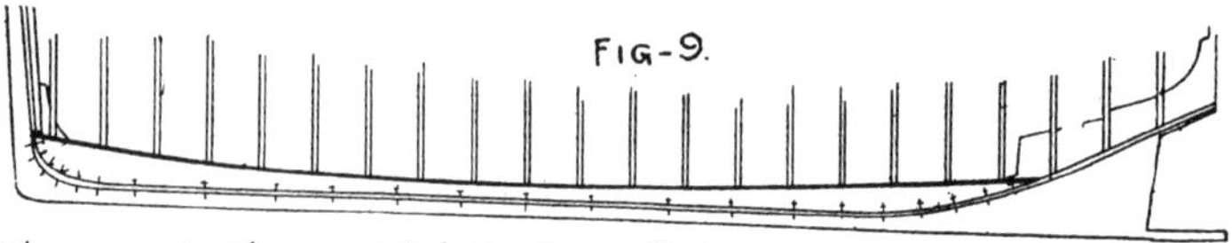

The process of spiling consists in bending a thin board roughly shaped to fit and then marking a series of spots with a pair of compasses set to a certain distance.

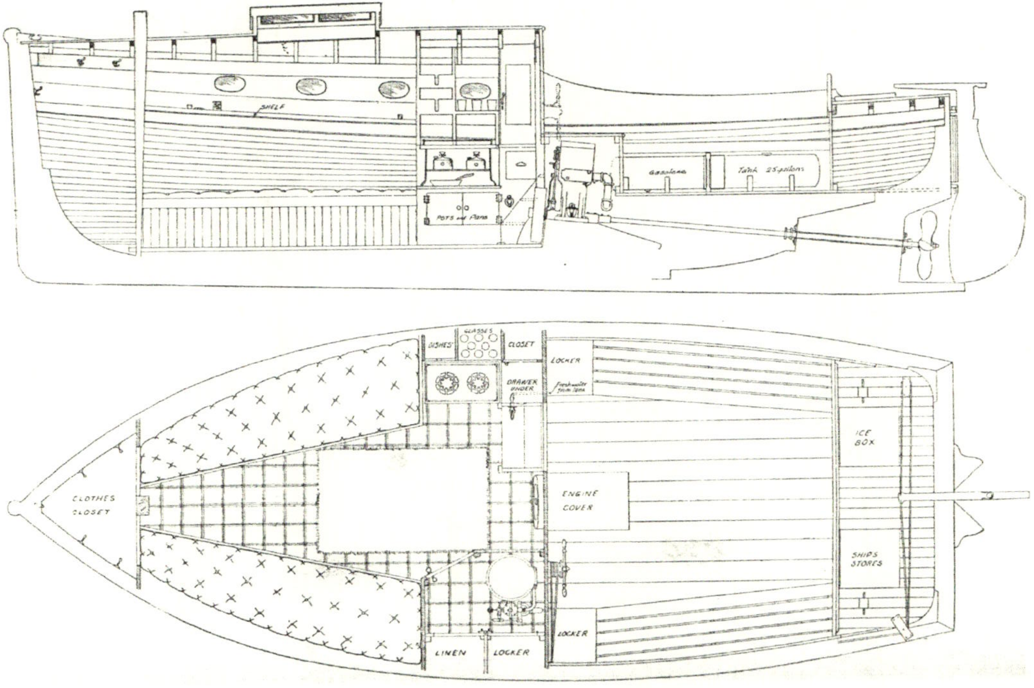

INTERIOR ARRANGEMENT OF "BEAVER"

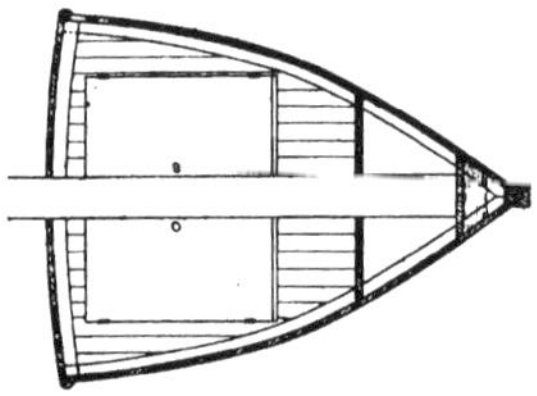

Ford end of cabin.

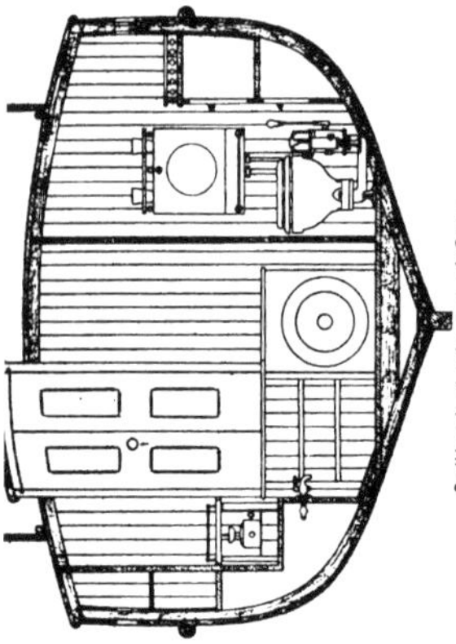

Bulkhead at after end of Cabin

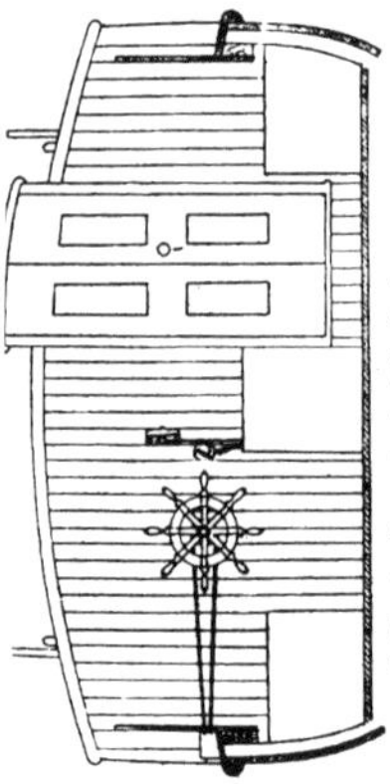

Bulkhead at forward end of Cockpit

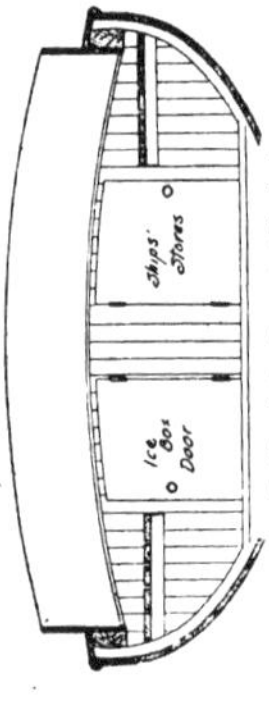

Bulkhead at after end of Cockpit

and lay it out flat on the wide, clear cedar board you have selected to make your garboard of, and proceed to set these distances back with the dividers still set to the same distance between their points. Draw a line through these spots with the aid of a thin batten, tacked outside of the spots, so the brad holes will not puncture the part you are going to use, and saw out the shape of the lower edge and plane it up smooth. The upper edge can be snapped with a chalk line perfectly straight, and cut out that way. If the boat is perfectly true on each side, this same pattern would fit both sides—you might try the garboard on the other side and see if it fits. If it does not, take another "spiling"; that is, repeat the measuring off process for the other. With screw clamps try these planks on, and mark any imperfections in fit, and cut until they fit perfectly all along the rabbet line. When perfect, you can fasten them on. For this you want two braces and bitts or twist drills, one with a bitt to cut for a countersink so as to let the nail head go in about 3/16-inch, so a wooden plug can be fitted, and be sure the bitt bores a hole the size of your plugs and not a sixteenth too large or too small, and another bitt that will bore a snug hole for your nails. The nails that go into the stem and deadwood should be galvanized iron nails, about 1¾ inches long, and those that go through the frames where you can get at them to rivet them up should be 3/16-inch copper nails 2½ inches long. Bore with a Dutch gimlet bitt for the fastenings into the ends, or you may buckle the nail over in the hard oak, and so split and spoil your plank. Above all, don't try to hurry the work—go carefully, and you'll get ahead faster in the end.

You can get out another plank or two to go above the garboard, making it a good wide plank, for when you get up around the turn of the bilge you cannot use them so wide.

Then with these two or three on the bottom put on what is called the "sheer strake," the one at the deck edge. With your thin pattern boards as a spile staff, find what shape the top edge of this board will make by measuring off at intervals with the dividers as before. Then make the plank about 6 inches wide in the middle, 4 inches at the bow, and 3 inches at the stern, sweeping in a fair line with a batten. Where the deck clamp comes in your way, so you cannot rivet the fastenings, use the galvanized iron boat nails instead.

Now you have to stop and do some figuring to see how many planks it will take to fill in the remaining space, so that you will not have a patchwork job. Take a batten and bend it down around the outside of a frame amidships. This distance, we will say, comes to 5 feet, and if you decide 4 inches is as wide as you want the planking to be there, it will require 15 planks. Don't make the common error of wanting to do the job quickly, and try to use planks a foot wide. Remember such a plank will shrink and swell nearly twice as much as one 6 inches wide, and not look well either; in fact, when you come to the turn of the bilge you may have to reduce them to 3 inches, putting in four planks 3 inches wide, instead of three of them 4 inches wide.

Now bend the batten around the second frame from the bow. Here you only have 3¾ feet, or 45 inches. Fifteen planks in a space of 45 inches gives you a width of 3 inches for each plank at that end, and the widths aft are found in the same way. In the same manner you could divide the distance at each frame and find out how wide the planks should be at every frame, but this is not necessary.

It is not always possible to get planks to run full length from end to end, and it is not necessary that they should, although most amateurs with their first smattering of knowledge on the subject imagine a well-built boat should have no butts at all. As a matter of fact, a well-made butt is the strongest part of the plank. Amateurs

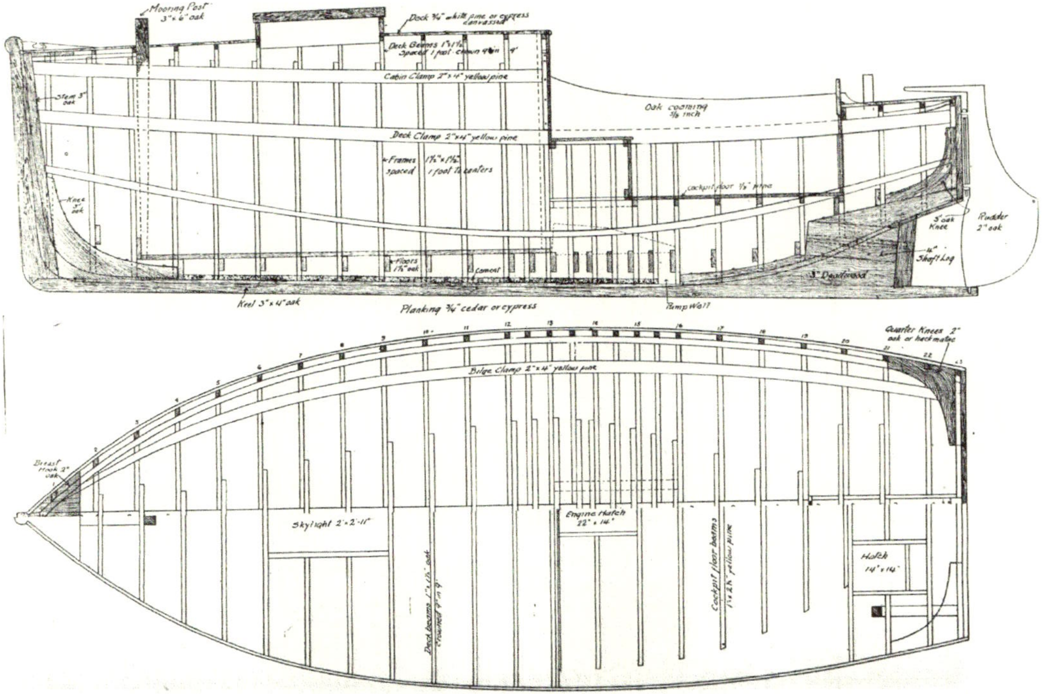

DETAILS OF CONSTRUCTION OF "BEAVER"

sometimes try to cut the planks so that the two ends both land on a frame with only a width of about 1½ inches. This is dead wrong. You can't toe-nail them both to the same frame and expect them to hold. Cut them so that they butt midway between the frames, and then fit a "butt block," as it is called—a piece of 1-inch or 1¼-inch oak plank, so it jams snug between the frames, and about ½-inch wider on each side than the planking, this gives a little ledge to extend up behind the planks above and below it, so as to steady it.

Do not let two butts come in line, one under the other. "Break butts," as boatbuilders call it, by making the joint in the next plank come two or three frames forward or aft of the first one. There should be at least two planks intervening between butts in the same frame space. With your "spile staff," find the shape the top of the next plank must be to fit the lower edge of the sheer strake, and with this shape marked off on a plank you are going to cut it out of, measure the widths 4 inches amidships, 3 inches forward and, say, 2 inches aft, whatever it may be, and sweep in a fair curve with a batten for the lower edge of the plank. After working three or four planks down from the top put some more on at the bottom, working toward the middle, until only one plank remains to be fitted. That is known as the "shutter" plank, and its fitting in in shipyards used to be the signal for a drink from the boatbuilders. Old timers would predict all kinds of disasters to the boat whose shutter was not "wet" to assist it in going in. The boatbuilders got the "wet." You may feel the joy that inspired this tradition when you realize that that plank completes the job of planking; anyway, you'll be happy and proud, too.

There used to be a great deal of mystery thrown about the job of planking a boat. It does call for some skill, but the foreman who laid out the planking always took good care to conceal the manner in which he did it.

The "planking scale" was a mystery in which none were to be initiated, and the lucky man who could pry into and understand the system at once became a power in a boat shop.

It is laughable how instructors in boatbuilding, when they come to a description of this subject fall down—as the author of one book I have in my library remarks, when he comes to describe planking: "I have never been able to find anyone who could explain this operation so as to make it clear, and doubt my own ability to do so, so will leave you to puzzle it out for yourself."

I don't want there to be any puzzle about it, and so I shall tell you here what the planking scale is, and tell you how to use it on this boat. For the scale plane up a thin slat of wood like a lath about an inch wide and ⅛-inch thick, and as long as the distance around the frame from the top strake to the garboard. To use this scale, butt one end against the top edge of the garboard or second strake, if it is on, and tack it lightly. Then bend the scale around the face of the midship frame, and mark where it touches the lower edge of the sheer strake. That is the distance to be planked, and as we have already decided our plank shall be 4 inches wide at this point, mark that spot 4. Then do the same forward, where we found the planks were to be 3 inches wide. Mark that distance as 3—the great mystery consists of dividing that distance between 3 and 4 into eight equal parts, and so making a scale on the slat of wood. Continue those same divisions up the scale to about 2, and you have a planking scale (Figure 10). On the top strake, at the second frame, mark two as the widths of all planks on that frame. To find the width on each frame butt one end of this scale on the top of the plank, on the bottom and where the lower edge of the top strake crosses the scaled off part of your batten or planking scale you can read the width of the plank from it. The widths so marked show how wide each plank is to be on each frame.

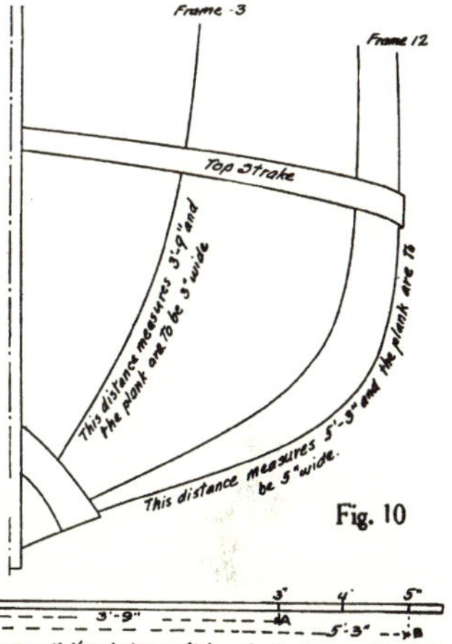

Fig. 10

Measure off the distance to be planked up with 5" plank on frame 12 and the same with 3" plank on frame 3. This gives spots B and A ... Let B represent 5 and A 3 on the plank scale - divide the distance between A and B in two parts and then each of these in eighths as each represents an inch. To find the width of the plank at any intermediate frame measure off with this scale by putting one end against garboard and where the underside of the top strake crosses scale read the width.

FIG-11

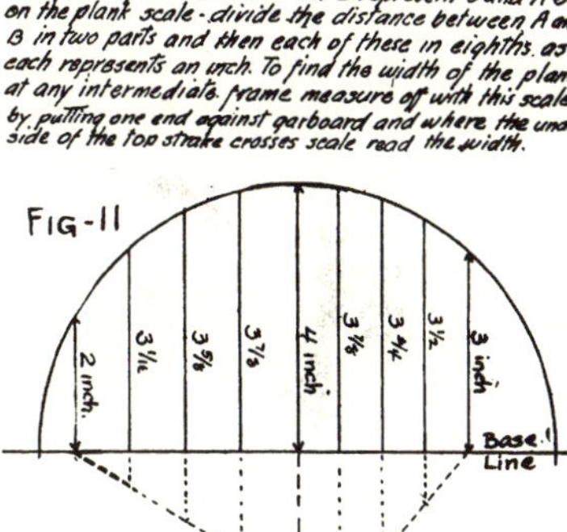

The dotted lines below the base line show a quick method of dividing the distances from the middle to each end width (the 3" and 2") If 4 parts are wanted, as drawn above, take any scale where four parts are longer than the distance to be divided and swing it down until the four mark comes to the center line

Another method for laying out a plank where you have determined the widths of the two ends and middle as we did at first is to strike a half circle, Figure 11, with a radius of 4 inches, the greatest width we decided for our plank. Measure up square to the base line to where the curve is 3 inches high—that is, the widths of the planks forward—and on the other side to where it is 2 inches. Divide the remaining space into any number of equal parts and you can measure the widths at these places and lay them out at corresponding intervals on the plank you are lining out on the board previous to cutting them out. This will give you a true, fair sweep for the other edge of the plank.

The three or four short strakes of plank to form the

Tapered bung reamer bit for cutting out knot holes.

raised deck forward will be easy enough to any one who can do the rest of the planking.

Each plank can be riveted up as it is put on. Punch the nail in solid then, with one man holding a heavy weight on against the head outside, another can get inside, put a copper "burr"—as the flat washers are called—on the nail, punching it on with a short piece of either brass or iron piping just big enough to go over the nail, cut it off within about 1/16-inch of the washer and rivet it up. Don't hit it a couple of smashing blows, as that will only buckle the nail in the wood. Take a light hammer with a ball pene end, like a machinist's hammer, and tap it all round the edge until it curls, or "burrs" over; then hit it a couple of good taps in the center to expand it.

Plane off the uneven seams and any hard spots that may show on the planking, then go carefully all over and test all loose looking knots, and punch them out. Those with a black ring, which is a sort of bark, are the loose ones.

Then you want a bung reamer, a tapered, gouge-shaped

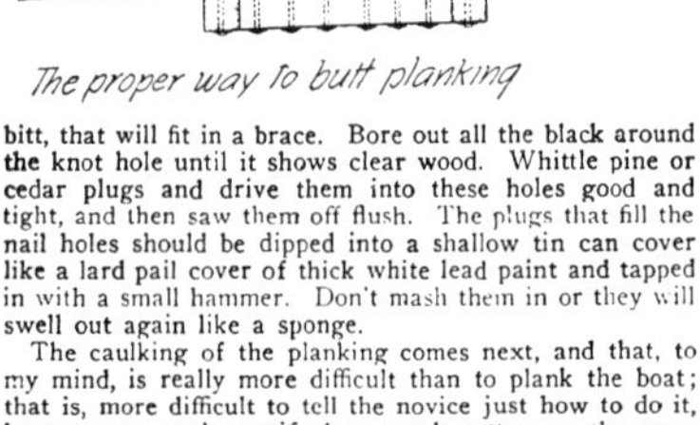

The proper way to butt planking

bitt, that will fit in a brace. Bore out all the black around the knot hole until it shows clear wood. Whittle pine or cedar plugs and drive them into these holes good and tight, and then saw them off flush. The plugs that fill the nail holes should be dipped into a shallow tin can cover like a lard pail cover of thick white lead paint and tapped in with a small hammer. Don't mash them in or they will swell out again like a sponge.

The caulking of the planking comes next, and that, to my mind, is really more difficult than to plank the boat; that is, more difficult to tell the novice just how to do it, because you can't specify how much cotton, as the amateur seams are apt to be uneven in widths, and require a little in one place and a lot in another. You want just to fil the seam up tight and yet not jamb the planks apart. I could tell better by sitting around and hearing the sound of the caulking mallet how well or how badly it was being done. The seams of the plank when put on should be slightly wedge-shaped, with the opening on the outside. Never fasten a plank on when the seam is wider inside than it is out. The water pressure is all from the outside, pushing on the cotton, and you want it to tighten as it is pushed in, and not to loosen.

The cotton should be driven in about a quarter of an inch beyond the surface of the plank and after painting with thin white lead paint over the cotton and allowing this paint to set over night, fill these seams with putty.

Then smooth off the plank with a plane, sandpaper it well across the grain and after painting over each knot with shellac to keep the sap in the knot from discoloring the paint, give the hull a prime coat of paint, either red lead or white lead mixed thin, and then two coats of whatever color you like.

If you want to do a nice job after you have once smoothed her all off, previous to painting, take a bucket of hot water and a big sponge and go all over her planking, soaking it well, and you will swell the grain and plugs, and if you then smooth her off again she will not become so rough due to the swelling when she is afloat for a few days. Full directions for striking the water line when painting will be found in the MOTOR BOAT HANDBOOK, VOL. I.

The framework for the cabin top, such as the openings for companionway slide and the skylight and the two hatches on the after deck, should then be fitted, and the beams for the cockpit floor fastened in place, but before these latter are secured, get out your motor foundation, notch it over the heavy oak floors, and bolt it solidly in place.

If you have your motor it is a good plan to line it up now, before you box the boat in with too many bulkheads and other things. Get a plumber to make you a lead sleeve to go through the deadwood and flange it over at each end and tack it fast into a good white lead bed on the faces of the deadwood. Get out the mooring post forward and the two towing posts aft, of sound, dry, seasoned oak. Locust is better, if you can get it. Fasten the two after ones in and brace them under the beams, but leave the mooring post, after fitting it, until you have laid the ¾-inch deck and stretched the canvas tightly over it.

First lay the deck, punch the nail heads in, plane down the seams, putty all the holes, and give it a good, thick coat of paint, and then stretch the canvas as tightly as you can. Pull it and tack it all around on the outer edge of the deck with copper tacks, and cover this edge, after it is painted, with a varnished half round oak molding.

After the mooring bitt is put through the hole in the deck turn the canvas up and tack it close around the bitt, and the same with the skylight hatch frame; screw it down in white lead onto the canvas, and then turn the canvas up around the inside of the hatch and tack it fast to prevent leaking.

It is necessary to hold the boat in some other manner when you come to finish the topsides, as the overhead shores are in the way, so nail a couple of short blocks on each side onto her plank, so the nails go into a frame and brace up from the floor to these with shores.

The oak covering boards around the edge of the after-deck are sawed out of 1-inch oak, and the deck laid of white pine planks 1 inch thick and 3 inches wide. Caulk and putty this deck and then plane, sandpaper and give it a coat of shellac.

For the coaming quartered oak should be used, or else mahogany, if you decide to finish her off in that wood. Fit the oak clock rails, as shown, forward and round off the head of the stem, so if you go alongside of a larger boat or up to a dock you do not have a sharp corner to cut and dig into things.

The main bulkhead at the after end of the cabin will have to be put up before you can put on the oak coamings

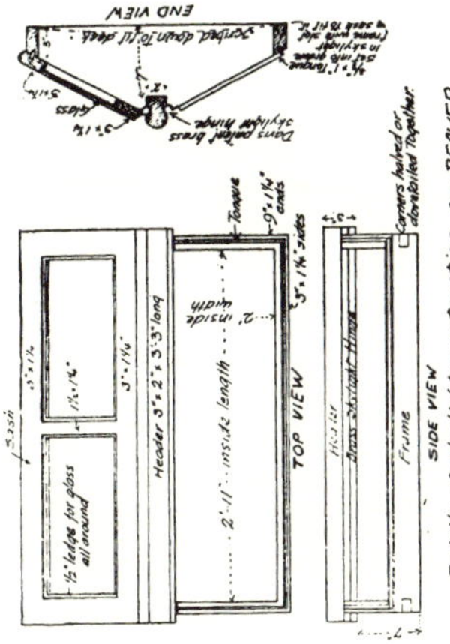

Details of skylight construction for BEAVER

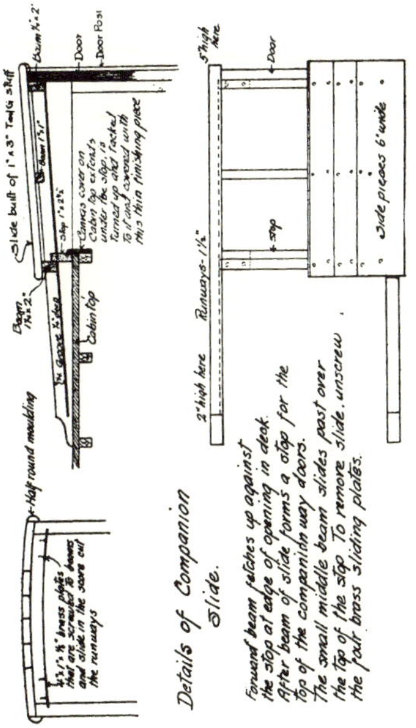

Details of Companion slide.

Forward beam latches up against the stop at edge of opening in deck. After beam of slide forms a stop for the top of the companion way doors. The small middle beam slides past over the top of the stop. To remove slide, unscrew the four brass sliding plates.

and finish the cockpit. In fact, this would be the first job to be done after the boat is planked, because you cannot even lay the cabin top until the up and down staving forming this bulkhead is complete. All this bulkhead stuff, both for here, the forward end of the cabin, and for the after end of the cockpit, can be of the same style of material, that is, either white pine or cypress tongue and groove staving, about 3 inches wide, with a bevel taken off each edge, so they form a narrow V groove when fitted together. This is an easy style to clean up and to paint. That is one reason why it is so largely used on boats in preference to any narrow scratch bead-work, which any one who has ever tried to clean up on a boat will fight shy of in the future.

The inside of the boat, from this main bulkhead to the stem, is ceiled up with the same material laid horizontal, beginning up under the clamp, and working down to the floor line. It is hardly necessary to tell one how to lay the cabin floor, or how to build the plain, boxlike partitions and cupboards which have been shown in the accompanying plans for the dishes, stove rack, etc.

The transoms are constructed by simply putting up a framework of about 1½ x 2-inch spruce, to give you the shape desired, nailing a corresponding cleat on the floor and then staving its sides up and down with the same kind of staving. Some amateurs may prefer to cut a wide, plain pine board and fit in her. It will do just as well, and perhaps look as well as the staving that we have shown. Make the seat tops of wide pine boards, so arranged that they can be taken up in sections, to get at the space below the cushions, to store provisions and duffle. The space forward of the mooring post has been bulkheaded off and fitted with two little doors to be used as a coat room, to hang up wet oilers, to lay sea boots away, and to hang heavy coats. We have not used this as a rope locker, as it is generally supposed to be used, but prefer to keep our cable coiled down on deck around the mooring post, and stopped to the deck with short pieces of line in small eye-bolts, where it will dry out and not rot.

There remain two difficult things to be done about the cabin top, and that is the construction of the companion way slide and doors, and the skylight. To assist the novice in constructing these I have shown detailed drawings of these two fittings. The subject of building the skylight alone is one that could fill a book; in fact, in my experience in running shipyards, I have had as many as six or eight boats a week come to the yard and their owners plead with me to come out and make their skylights tight. Any one who has done any amount of boating knows how disagreeable a leaky skylight can become in wet weather. For that reason be very careful in constructing this one, to make all your joints tight, and try and get a little comfort in *Beaver*. The companionway slide is comparatively simple, when you study out the accompanying detailed plan.

Oak chocks or saddles are to be fastened to the cockpit floor where your seats are to be built, to accommodate the cylindrical tanks, one on either side, one for fresh water and the other for gasolene, as shown in the plans. When these are in place, fasten the cleat that holds the after end of the seat across that bulkhead, and on the after side of the little square locker in the forward end of the cockpit put a corresponding cleat. Then build your seats of the long, narrow slats as shown in the plan, held together by cleats underneath these slats, but do not nail them fast at the ends. Leave them so you can lift this slat seat right up out of place, so that you can get at your tank whenever necessary by taking out a couple of screws.

On the after-deck get out two square oak frames of 1-inch square stuff, the size of the hatches you have formed in the deck there and build two square covers out of about ¾-inch stuff, and for tightness sake, cover them with canvas, held around the edges with a small half-round oak molding and paint them.

We have made no attempt in *Beaver* to make the cockpit floor a water-tight one, because the floor level is only a few inches above her load waterline, but if one prefers, he could caulk it, and put lead pipe scuppers in the after corners, but if one does this, I would advise raising the cockpit a few inches higher than shown in the plans, so there would be no danger of the water coming back through the scupper when the boat settles by the stern when running hard.

The steering gear of *Beaver* is made as simple as it is possible to make it, and is all get-at-able in case of a breakdown. Sheaves set in oak chocks are bolted to the deck, as shown, and the tiller ropes lead outside the coaming on the port side, then in and over her steering wheel, which is bolted to the after bulkhead between the box over her engine and the locker on the side which enables a man to steer either right-handed or left-handed.

It is hardly necessary to go further into the details of building such a boat, as they are all of such minor importance that even a boy would know enough to go ahead and complete the job, and we feel we have fully explained the difficult parts, where the amateur would need some help.

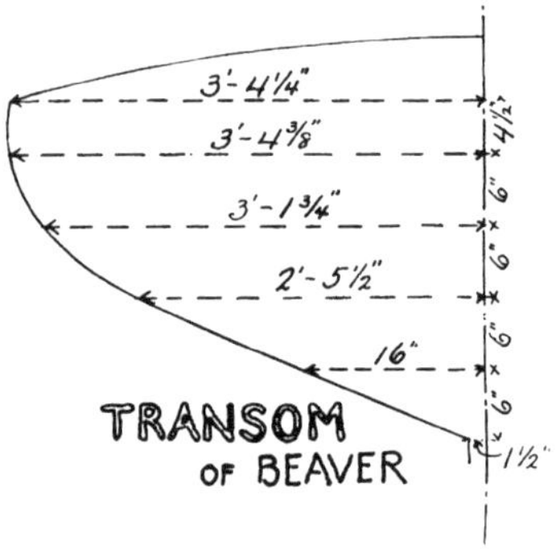

TRANSOM OF BEAVER

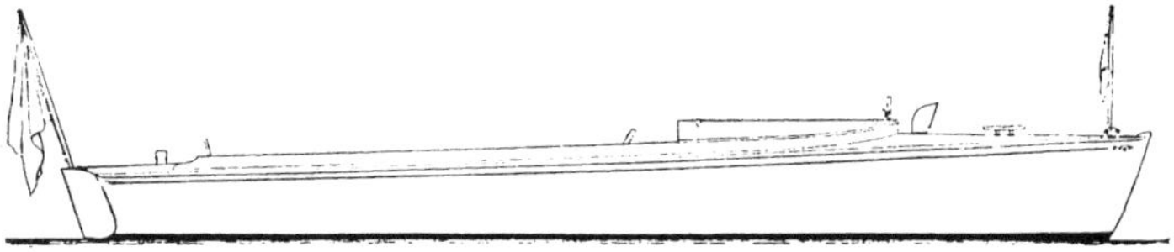

Outboard Profile of the Nock Runabout

How to Build a 25-Foot Runabout

PART I. BY FREDERIC S. NOCK

DIMENSIONS.

Length over all	25 feet,	9½ inches.
Length, waterline	24 feet,	0 inches.
Breadth, extreme	5 feet,	0 inches.
Breadth, waterline	4 feet,	5½ inches.
Draft to Rabbet	0 feet,	11¾ inches.

THERE seems to be quite a demand for a small runabout, something a man with some knowledge of tools can build, and, when completed view his handiwork with pride. Not only is there a great deal of pleasure to be had running around in one of these craft, but there is a certain charm and fascination in building one, at least many men view it in that light. That such a diversion should appeal to the younger boys is rather to be expected, but as I look around and find many of the older boys who spend the day in their office, devoting all their leisure moments building a boat in some barn or shed, I cease to be surprised at the demand for articles on the building of small boats, even in spite of the great number of plans that are illustrated and described in the different magazines devoted to the sport.

The building of boats by amateurs is by no means confined to very small craft, as I can recall to mind four boats that are being built in the vicinity of Providence, by amateurs at the present time, all of them being over 32 feet in length. Some of them have been started recently, others were started long ago, and some are far from being finished at the present time. With one exception, the boats are being built throughout by the owners with assistance, at times, of their friends. The exception had the frame gotten out for him and shipped K. D., which means knocked down or taken apart before being shipped. Another point which shows a decided step in the right direction is that the boats in question were all designed by competent men, and are not a sample of the old rule of thumb type.

The average man who wants a 22 or 25-footer is by no means satisfied to consider a boat with a speed of 6 to 7 miles an hour, even though he wants a family boat. A few years ago when a man wanted plans for a family boat it was supposed to be something very wide and comfortable, that looked as though it would be impossible to capsize, and, needless to say, it was usually very slow. Nowadays when a man wants a family boat of 25 feet in length he is quite apt to give the designer a decided jolt by stating that he must have a speed of 12 to 16 miles, and he is pretty apt to get it. I don't mean to infer by this that he has a boat of the length mentioned, carries a Sunday School class in the boat, and drives her at the required speed, but he can have a boat that with three or four persons on board can make the required speed and not require an engine of excessive horsepower. At the same time the boat would be capable of accommodating some ten or more persons with perfect safety. The speed mentioned is not by any means the limit of a 25-foot boat as 24 or more miles an hour can be obtained, but the power plant is usually something that plays quite an important part, and you cannot very well expect to obtain such a high rate of speed with a 10 or 20-hp. engine, no matter how light an engine you may procure.

The boat I am going to call your attention to, and try to explain how to construct, is designed to carry a small engine weighing between 600 and 650 pounds. You can procure an engine of from 5 to 50 hp. that will not exceed the weight in question, and thus the power question ought to suit the most exacting.

The plans show the boat without a skeg to protect the propeller, etc., but I have also shown the same keel with a shoe and skeg for those who may prefer same. If you are looking for speed, don't use the skeg as it simply means increased wetted surface and consequently more resistance.

Many of you fully understand how to "lay down" the lines, but for the benefit of those who are not initiated into the mysteries and want to start at the beginning, I will try and explain the operation in as simple a manner as I know how. In the first place it means the reproduction of the lines full size on the floor, or whatever you draw them upon. I shall have to presume that you have a barn or suitable building in which you intend to build this boat; if not, and you have to rent one, select a place that has a fairly smooth floor and plenty of light. Clear a place on the floor about 27 feet in length, and five or six feet in width. If the building does not possess a floor or it is in poor condition, you can make up a board large enough to draw the lines on, using spruce or hemlock boards, cleated together on the underside.

It is advisable to work from the base line, and this you can put in near the edge of the board, make a mark at either end, then with a chalk line stretched taut and snapped, you should have a straight line through the two points in question, but as chalk is very easily erased it is advisable to mark over this line with a pencil. Select a board that is true and straight along the edge, and use this for a straight edge to draw in the pencil line. This line is the base line and all the heights given on the laying down tables are above the base line. Parallel to this line and two feet above same, strike in another line, mark over it with pencil to prevent it being easily erased. This corresponds to L. W. L. (load waterline) on the plan. Above this line (L. W. L.) at intervals of

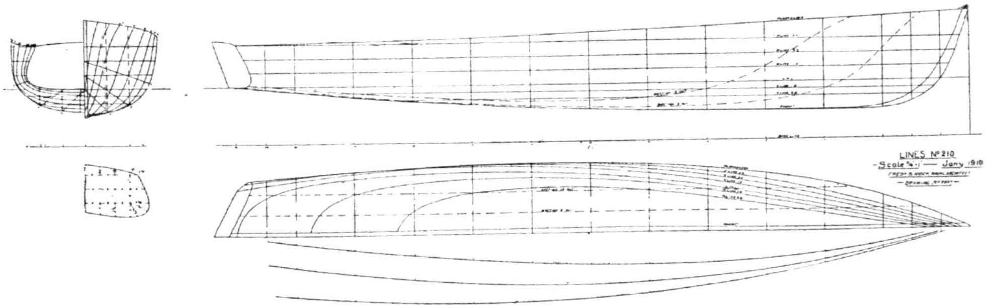

Fig. 1. Lines of the Nock Runabout

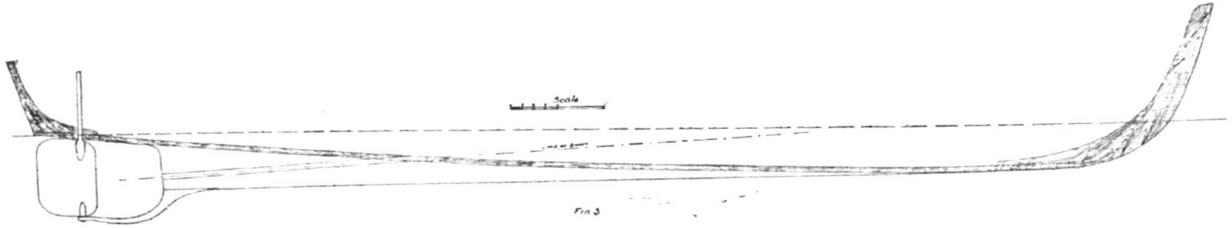

Fig. 3. Showing Keel, with Skeg and Shoe to Protect the Propeller

5 inches, you can draw three more lines parallel to the base-line and counting from the L. W L. mark them W lines, 1, 2, and 3, above, which means above the L. W. L. There are still two more lines to be drawn parallel to those mentioned, and they are W. line 1B and W line 2B. W. line 1B is 3 inches below the L. W. L. and 2B is 3 inches below 1B Draw these lines in the same manner as the others, and mark them in their proper order so that you will not make any mistakes. Begin at the right hand side of the baseline and with a square draw a line perpendicular to the base line, and about 4½ feet in height, measure 1 foot to the left and draw another line parallel to the first, then at intervals of 2 feet draw 12 more lines all perpendicular to the base line, and, of course, parallel to each other

Beginning with the second one from the right, mark this No. 0, the next No. 2, the next No. 4, and so on corresponding to the stations on the plans, and perhaps it would be as well while you are putting in these lines to draw another line between stations 0 and 2, and also 22, and 24, numbering them 1 and 23 respectively, then you will have just as many stations on the board as is shown on the plans. Now if you will consult the laying down tables, you will note that the upper line of the table proper gives the heights (above the base line) of the sheer at the different stations, mark these points on the perpendicular lines, taking great care to have the right measurement. Drive a small bung head wire nail at each point, then take a batten about ¾ inch square and 27 feet long, one made from a good, clear, straight grained piece of stock—yellow pine is as good as anything as it will bend fair and true—spring this batten up to the nails, marking the points and keep the batten in position by driving nails in the board on the opposite side of the batten. Sight carefully along the edge and see that it is a nice, fair curve; if there are any short kinks in it move the batten until it shows fair and true, and then with your pencil draw in the sheerline, which represents the height to the top of the phanksheer.

The next line shown on the tables is the rabbet line. You can proceed to put in the points for this line in the same manner as you did the points for the sheer, and when your points are all marked, spring your batten to same and mark the rabbet line. You will undoubtedly find that the batten you have been using for the sheer-line is too strong to make the curve at the forward end of the rabbet line where the stem connects with the keel, and, therefore, in order to put this in it will be found necessary to have a much lighter batten. Procure one about 3-16x½ inch and if you get one about 5 feet long, you can use it for the curves of the body sections. In order to complete the rabbet line to the sheerline, you will have to mark the distance the rabbet line is from the point of measurement on the plans, and detail of stem. Mark each of these and continue the line to the sheer, taking due care to have it fair and you can proceed in a similar manner to put in the line representing the keel bottom, taking the measurements from the tables. The face of the stem is a continuation of the keel bottom and you can get the measurements of same from the plans, or the separate drawing of the stem. Draw this line in the same manner as you did the rabbet line. If you want to put in sections 9 inches and 18 inches out, you can do so, although you will not derive a great deal of benefit from same.

The next thing is the waterlines. The table shows the half breadths at the different stations, beginning with the sheer. As the amount of space you have is limited, you can use the baseline as the centerline and mark in the waterlines in practically the same manner as you

LAYING DOWN TABLES FOR LINES Nº 210

All Dimensions given in Feet, Inches, and Eights, and Above Base Line, 2 Feet below the L.W.L.

All Lines to outside of Plank

	Stations	0	1	2	4	6	8	10	12	14	16	18	20	22	23	24
HEIGHTS	Sheer Line	4.4.2	4.3.6	4.3.1	4.1.7	4.0.6	3.11.6	3.10.6	3.9.7	3.9.0	3.8.2	3.7.5	3.7.0	3.6.5	3.6.3	3.6.1
	Rabbet	2.7.5	1.6.3	1.1.6	1.0.2	1.0.2	1.0.4	1.0.7	1.1.5	1.2.5	1.3.6	1.5.5	1.7.3	1.9.5	1.10.6	2.0.0
	Keel Bottom	2.0.0	1.3.2	1.0.4	1.0.0	1.0.0	1.0.2	1.0.6	1.1.5	1.2.5	1.3.6	1.5.5	1.7.3	1.9.5	1.10.6	2.0.0
	Section 9" out		4.0.5	3.1.1	1.8.6	1.4.0	1.2.3	1.2.1	1.2.3	1.3.2	1.4.3	1.5.7	1.7.6	1.9.7	1.11.0	
	" 18" "				3.8.7	2.6.2	1.9.2	1.6.1	1.5.2	1.5.2	1.6.0	1.7.4	1.9.2	1.11.1	2.0.2	
HALF BREADTHS	Sheer Line	0.5.6	0.10.5	1.2.5	1.8.6	2.0.6	2.3.3	2.5.0	2.5.7	2.6.1	2.5.5	2.4.3	2.2.2	1.11.2	1.9.4	1.7.5
	W Line 3 Above	0.2.7	0.6.6	0.10.2	1.4.6	1.10.2	2.1.7	2.4.2	2.5.5	2.6.0	2.5.5	2.4.3	2.2.3	1.11.3	1.9.5	
	" 2 "	0.2.0	0.5.3	0.8.6	1.2.6	1.8.0	2.0.1	2.3.0	2.4.6	2.5.5	2.5.5	2.4.6	2.3.2	2.0.7	1.11.4	
	" 1 "		0.4.1	0.7.1	1.0.6	1.5.7	1.10.1	2.1.2	2.3.4	2.4.6	2.5.0	2.4.4	2.3.2	2.1.3	2.0.2	
	L.W.L		0.2.7	0.5.5	0.10.6	1.3.3	1.7.4	1.10.6	2.1.0	2.2.2	2.2.5	2.2.0	2.0.1	1.8.5	1.4.4	
	W. Line 1 Below		0.2.0	0.4.3	0.9.2	1.1.6	1.5.7	1.9.0	1.11.1	1.11.7	1.11.5	1.10.0	1.5.6			
	" 2 "			0.3.1	0.7.1	0.11.2	1.2.7	1.5.6	1.7.4	1.7.4	1.5.6	0.10.3				
Diagonal	A		0.3.1	0.5.6	0.9.2	0.11.6	1.1.2	1.1.6	1.1.4	1.0.4	0.11.0	0.8.6	0.6.3	0.3.4	0.1.6	
"	B		0.4.4	0.7.2	1.0.3	1.4.3	1.7.2	1.9.0	1.9.7	1.9.7	1.9.0	1.7.2	1.4.6	1.1.6	1.0.0	
"	C	0.2.2	0.5.2	0.8.2	1.2.0	1.6.7	1.11.0	2.2.0	2.4.0	2.5.1	2.5.4	2.5.0	2.3.6	2.2.0	2.0.7	

Diagonal A Intersects Perpendicular 2'-0" Above Base Line and Base Line 2'-0" out

" B " " 2'-6" " " " " " " 3'-6" "

" C " " 3'-0" " " " " L.W.L 2'-9" "

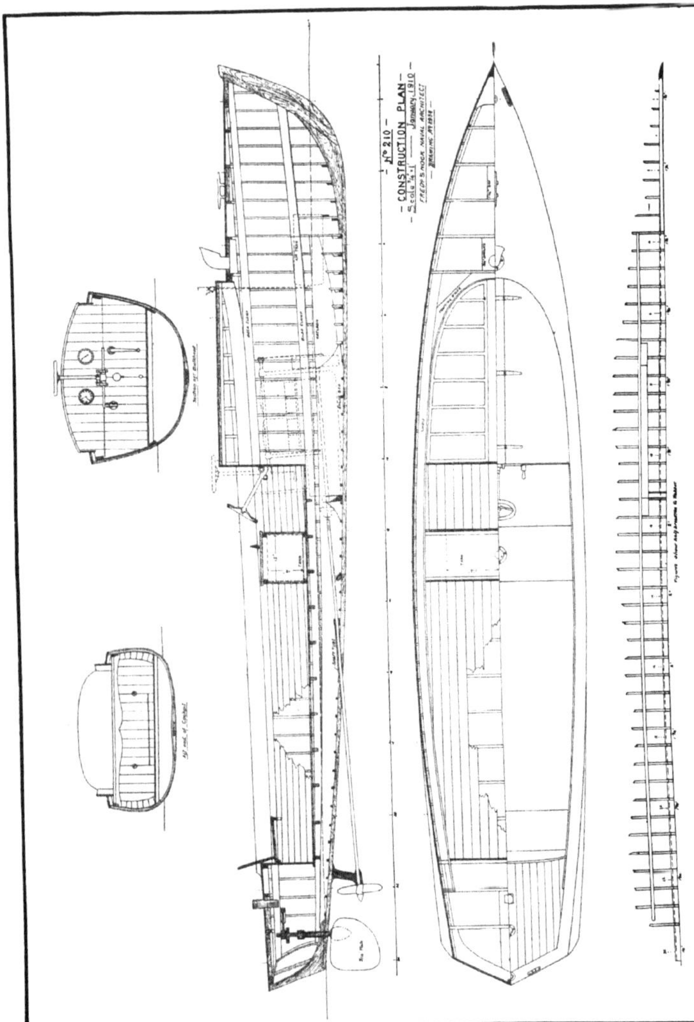

Fig. 2. Construction Plans

did the rabbet line on the elevation, using the strong batten where possible and the lighter batten for the short curves. Be sure and see that all the curves are fair before you draw the line, and in doing this, while it takes more time, it will save you lots of trouble and worry later on.

The waterlines do not come to a point at the stem, being 3-16 inch out from the centerline. This applies to all the waterlines and if you will bring them to this point, you will find the advantage of doing so when you start to make the stem, as each of these lines can be referred to. The body sections should next claim your attention. Use either of the center stations, preferably No. 12, as the centerline and put in the diagonal lines A, B, and C, as shown on the plans. The tables show that diagonal A intersects the perpendicular 2 feet above the baseline and the baseline 2 feet out, therefore, you can measure up on the centerline 2 feet and you will find that the L. W. L. crosses at that point. Measure out on the baseline 2 feet either side of the centerline and then draw a line through the points on either side and you have diagonal A, Diagonal B intersects perpendicular 2 feet 6 inches above baseline and baseline 3 feet 6 inches out. The point on the centerline will be where W. line 1A crosses the perpendicular. Measure out 3 feet 6 inches along the baseline either side of the centerline, draw a line through the points on either side and you have diagonal B. Diagonal C intersects the perpendicular 3 feet above the baseline and the L.W.L. 2 feet 9 inches out. The point of intersection with the centerline is where W. line 2B crosses same, and if you mark a point on the waterline the required distance out and then draw a line through the points, you have diagonal C.

It is advantageous to work to the measurements on the diagonals wherever possible for most of them cut the body sections at a less acute angle than the waterlines, especially the lower ones. Where the waterlines or any line intersects another at an acute angle, there is liable to be a chance of making a slight variation in the measurement.

Referring to the tables, you will note the half breadths as given for the sheer, and waterlines. Start with section No. 12, mark off the points on the different waterlines, then the diagonals. The height of sheer is already marked on the board so that all you will have to do will be to get the half breadth of the sheer and mark this on the point representing the height. There is still another point required before you can draw the section and that is the one representing the rabbet line. You already have the height marked on the board. Draw a line at right angles to the centerline cutting the rabbet-line on station No. 12, and then refer to the half breadth plan of the keel for the width of the rabbet at this section and mark this point. You can also draw in two lines on either side of the centerline, one 9 inches out, the other 18 inches out. These lines represent the sections 9 and 18 inches out, in a line fore and aft, and used to be called bow and buttock lines. By using these lines and marking on same the heights at different stations, it will allow of your having a greater number of points

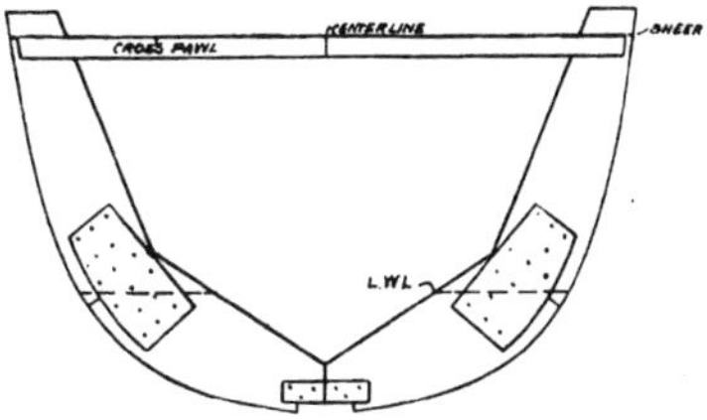

Fig. 6. Mold, Showing Cleats and Cross Pawl.

to draw your line through that which represents the body section. Use the light batten, bend to the points, keeping the batten in position in the same manner as you did for the rabbet and waterlines. Ascertain that it is fair and true and draw the line.

You must bear in mind that the lines as laid down are to the outside of the planking and if you prefer to lay down the body sections the molded size, or the size of the inside of the plank, you will have to take off 7-16 inch (the thickness of the plank) all around the inside of the curve. You can please yourself whether you lay down the lines as shown, or draw them in to the molded size. If you prefer the latter method, you can do so by setting a pair of compasses to the required dimensions, 7-16 inch, set the spur leg against the inside of the batten and with the pencil end, draw the line 7-16 inch inside of the batten; take particular care to keep the spacing an equal distance away from the batten the entire length of the curve.

Draw in all the body sections you require, and you can proceed to finish the lines, for while you have completed them as far as the lines shown on the plan are concerned, you still lack some very important lines, such as the inside of the stem, the upper side of the keel, transom knee, line of shaft, etc. These you can obtain from the construction plan and the detail drawing of the stem, transom, etc. It is also advisable to draw in the scarf of the stem and keel, also the transom knee.

Having completed the drawing to your satisfaction, you can proceed to get out the molds. Hemlock, spruce, pine, or any such material that is not expensive is suitable. If you want to make molds for each of the sections you will require about 110 feet of ⅞-inch stock. The prac-

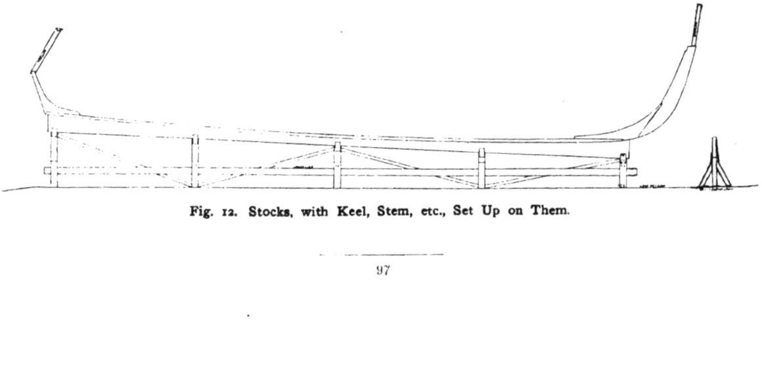

Fig. 12. Stocks, with Keel, Stem, etc., Set Up on Them.

tical builder is not apt to space the molds every two feet apart, but would in all probability use molds Nos. 1, 4, 8, 12, 16, 20 and 23, setting them in their proper places. If you adopt this method, you had better use battens a trifle heavier than you would use if your molds were spaced 2 feet apart.

Take some small wire nails, with large heads, lay the head of the nails on the lines, points directed toward the center of the body sections. These nails should be placed at intervals of 2½ to 3 inches, tap them with a hammer to hold them in position and then on top of same press one of the pieces of wood you intend to use for a mold. When you raise the piece you will note that there are a number of small dents made by the nail heads. These represent the points of the curve, and if you use your batten and draw a line through the points, you have a reproduction of the mold. Saw to the line, fair with a plane or spokeshave and then mark on another piece of board a duplicate of same, and you will have the two halves of one of the molds. You should bear in mind that you will need some marks to go by when setting up these molds, therefore, it is advisable to draw the L. W. L. and the sheerline on same. If you have the use of a band or jig saw, you can save time by fastening the first piece of wood, with the shape marked thereon, to another piece, and saw them both out at the same time.

Fasten the mold together, using a piece of stock about 2 inches wide for the cross pawl, and as a guide you would do well to make the upper edge of the cross pawls on a line with the mark representing the sheer. Carefully measure the completed mold in order to ascertain that the widths at the sheer and waterlines are correct. If you cannot get stock wide enough to make the halves of each mold in one piece, you will have to join them, and the simplest method of doing this is to butt the joints together and cleat them. (See Fig. No. 6.)

Place the cleats well in from the edges, and nail a piece of stock across the lower part of molds in order to have something to fasten them to the keel. Make all the molds up, measuring each one carefully to see that they are exactly the same dimensions as the full-size drawing.

You may begin to think that there is a lot of unnecessary measuring, marking of lines across the molds, etc., but I can assure you that if you have a number of points, it will help you check the molds when they are in position and show the least error.

The practical man can tell at a glance whether the molds are all fair, but this does not always apply to the amateur. After you have completed your molds, you can proceed to get out the stem. It is to be of oak or hackmatack and if you can secure a natural crook large enough to make the stem and knee in one piece it is to your advantage to do so, otherwise you will have to make it up of two pieces as per plans.

The specifications call for the stem to be sided 2½ inches, (or, in other words, it is to be 2½ inches from one side to the other, not from the face to the after side). It is a good plan to get out a template of the stem, and mark on same the rabbet line, and then after cutting the stem to the required shape you can mark on one side the shape of the rabbet, reverse the template and mark the rabbet on the opposite side. One of the simplest methods of doing this is to bore a series of holes through the rabbet line on the template and then with a drill or awl reproduce these points on the stem, bend a batten and draw a line through these points and you have the rabbet line.

If you make the stem up of two pieces fasten them together with 5-16-inch bolts, heads to be riveted over washers, or you can draw it together with nuts and bolts if you prefer. The heads should be well countersunk and covered with wood plugs.

Draw a centerline down the face of the stem and 3-16 inch on either side draw lines. Fasten the stem in some manner as in a vise and then proceed to trim from the rabbetline to the line 3-16 inch from the centerline of

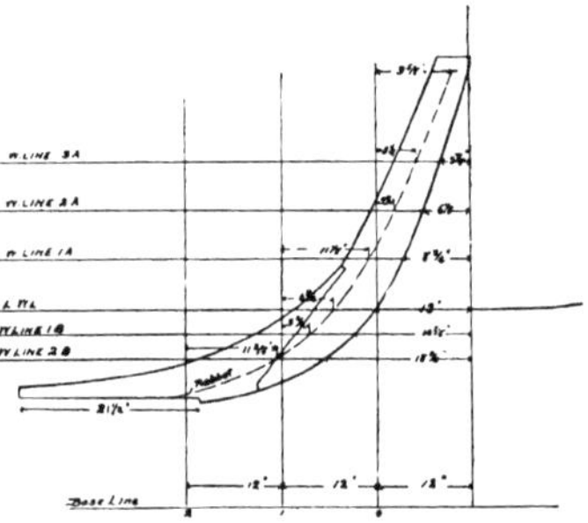

Fig. 5. Stem, with Dimensions Marked Thereon

face. Continue to where waterline 2B crosses the stem, trim both sides and this will leave the face of the stem ⅜ inches in width. Below the point mentioned (intersection of W. line 2B with face of stem) the width of the face increases until it is about 1⅛ inches wide where the stem or knee joins the forward end of the keel. Fasten the stem down to your bench so as to allow you to get at it handily and proceed to cut the rabbet. A fid is used by some builders as a template for cutting the rabbet to the required depth and bevel. As the bevels on the sides of the stem of this craft are fair with the plank to a point just below waterline 2B you can easily see how the rabbet would require to be cut by trying a piece of wood of the thickness of the plank, 7-16 inch, and cutting in the stem until it sets flush with the side of stem at the rabbet line, and on the same bevel as the sides of the stem forward of the rabbet line. If you want to get this down to the fine point, you will have to refer to the full-size drawing of the boat. You will note that the plank at the sheer intersects the stem at a certain angle and that the intersection of each of the water lines is of a different angle. Make a template showing these bevels and then cut out the rabbet at these points to correspond to the bevels you have taken, them trim and fair the intervening spaces and you will find that your rabbet is cut correctly. It is not a good plan to carry your rabbet clear to the lower end of the stem or knee until it is fastened to the keel, for by finishing it afterwards you can be more sure of it being fair.

When you have finished the stem set it to one side and get out the keel. Select a good piece of oak about

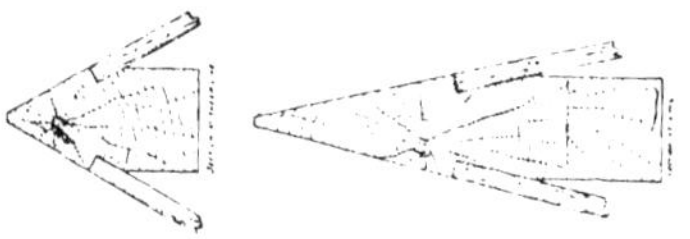

Fig. 8. Section Through Stem at Sheer and at L.W.L.

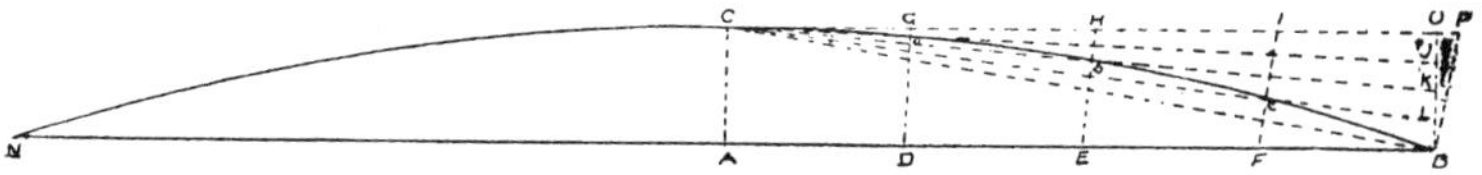

Fig. 10. Showing Method of Constructing Curve for Beams

24 feet in length, 6 inches wide, planed to 1¼ inches in thickness. Draw a line through the center and then with a square mark the position of the different stations. Figure 4 shows the half breadths to the rabbet of the keel at the different stations, set these off on either side of the center-line, then spring a batten to the different points and draw the line, taking care that the line cuts the points. You will then have the shape of the underside of the keel and as the upperside is ⅝ inches wider than the rabbet line on either side, you can mark same, and cut to this line, using a band saw if it is available. Smooth the edges and then cut out the rabbet, which is ⅝ of an inch up. If you can get this cut on a circular saw or molding machine, it will save you time, as it is rather a tedious job chipping it out with a chisel and mallet. The rabbet at the after end and well forward of station No. 12 is practically square, but it begins to have a decided bevel the nearer you get to the stem. If you want to know just how much this bevel is at each of the stations you can mark out on the full-sized drawing the thickness of the plank next to the keel and take the bevels off same. The upper edge of the keel is to be well rounded on the edges as per Figure 9 and in doing this it will enable you to make the frames bend nice and fair where they cross the keel. Cut the keel to the required length, shape the after end to fit the transom knee, the forward end to

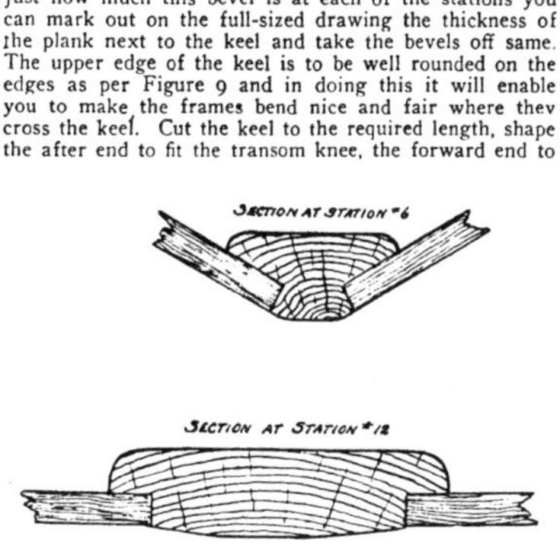

Fig. 9. Section Through Keel at Stations No. 6 and 12

fit the stem, and the keel is ready for setting up on the form.

The knee for the transom is to be of oak or hackmatack, preferably the latter, as it is lighter. It is to be sided three inches and cut to shape as shown on plans: Draw a line through the center of the after side and trim to the same bevel as the transom, and then on either side of the centerline ½ inch out, draw another line which will represent the rabbet line; on either side of the knee parallel to the face draw a line the thickness of the plank for transom and cut the rabbet, (see Figure 11). This knee, when you have finished the rabbet for the transom, can be fastened to the keel with 5-16-inch bolts riveted over washers. Take care that you do not get any bolts through the knee where they would be likely to interfere with the rudder port. Before proceeding any further with the framework for the boat it would be as well to make up the stocks to hold the boat while being constructed. For the form of the keel you can take a piece of spruce plank 2 inches in thickness, about 10 inches wide and 25 feet in length. Mark on same the shape of the keel bottom, and the stations, saw to the line, square the edge where the keel will set on same, and set this form up on posts made of 2x3 or 2x4-inch spruce.

The bottom of the keel should be about 2½ feet above the floor so that in planking the boat, you will have ample room to work under it. Strike a line on the floor and set the posts to this line, at either end of the form cutting the posts at such a height that the distance from the floor to the upper edge of the form at stations Nos. 2 and 23 are proportionately the same height above the floor as they would be above the baseline.

Perhaps it will simplify matters for you if you make a mark on these posts representing the baseline and measure up from same to the top of the mold and the height at the different stations would be the same as is given in the table of heights above baseline to keel bottom. Of course, you must make sure that the line representing the baseline is perfectly level irrespective of the floor as you will undoubtedly work from the line.

About 5 posts in all would be enough to support this boat. Brace them in both directions so that they will be perfectly rigid. (See Fig. 12.) Sight along the upper side of mold to ascertain that it is straight and you can proceed to get the keel ready to fasten to same. Fasten the stem to the keel with 5-16-inch bolts riveted over washers. The keel can be fastened to the form with long screws placed about 2 feet apart. Put washers under the screw heads and this will enable you to draw the keel down to the shape of the form without pulling the heads into the wood. Plumb the stem and with staylaths fasten to the rafters in order to hold it in position, plumb the transom knee and secure by staylaths in the same manner as the stem. Before you proceed any further, it would be advisable to check the keel, etc., in order to ascertain whether the centerline is straight.

We will assume that you have set the stem and the transom knee plumb, so you can take an awl or small nail, drive it in the centerline at the head of the stem and another in the centerline at the head of the transom knee, stretch a strong line or piano wire from one to the other, then take your plumb bob and drop a line from same. If the point of the bob intersects the centerline of the keel at either end or in fact anywhere along the line you can rest assured that the centerline of the keel is straight.

If you have laid out on the floor the centerline of the shaft you can now transfer it to the keel and bore the hole as it can be done more readily before the molds or frames are in place. Having finished boring the hole you can turn your attention to the finishing of the rabbet-line on the lower end of the stem where it connects with the keel, and unless you have already done so, it is well to fair the rabbetline at the forward end of the keel. The forward sections intersect the keel at an acute angle, which diminishes as you work aft. The bevel of the rabbetline at any station can be taken from the lines laid down on the floor. Some builders will not finish trimming the rabbet on keel until the molds are in position and use them as a guide for the bevels, but you will not save any time by adopting this method. The transom should be gotten out and fastened to the knee before you set up the molds. Select a nice clear piece of white pine ⅝ inch thick, large enough to make both sides.

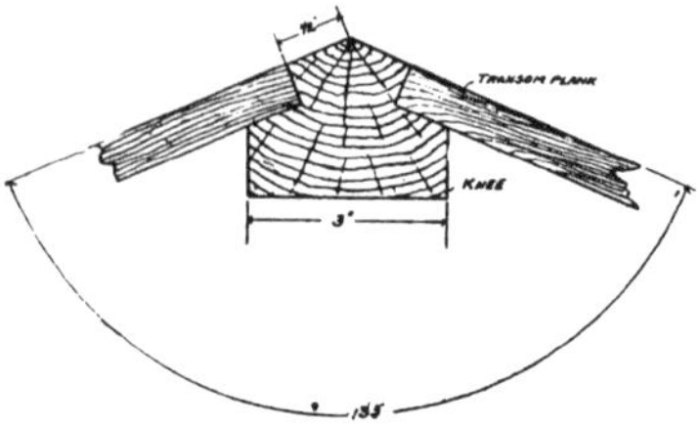

Fig. 11. Section Through Knee, Showing Angle

Mark out the shape as per plan showing half of the transom (Fig. 13) cut out two pieces, one for either side. The dimensions given on the drawing are taken from the center of the knee, and as there is a half inch of stock on either side of the knee where the transom plank sets into the rabbet, you will have to take off a half inch from both of the pieces. Carefully trim and fair the edges, then mark on the inside of both pieces the shape of the inside of the plank at this point. Now mark another line around the edge where the plank will be fastened to same, this line to be about ⅛ inch in from the afterside. Cut from the line on the forward side of these pieces to the line on the edge. By having the ⅛ inch of stock at the after edge, you will have plenty of material to calk to, when you calk the ends of the plank, and will not have the inconvenience of breaking out little pieces along the edge as so often happens with the amateur if the edges of the transom are brought to a knife edge.

The frame, or check pieces, which are to be fastened to the forward side of the transom, should be made of oak about 1¼ inches wide and ⅞ inches thick. They need not be in one piece and can be made up in sections if you desire. See that they are trimmed to the right shape and the edges bevelled to the right angle so that the inner side of the plank will set fair against same. The bevels can be obtained from the full-size drawing, by taking the angle at which the waterlines meet the transom. Fasten the cheek pieces securely in position and then you can fasten the two pieces of the transom plank to the knee. Ascertain that they are set in the right position, and then measure out from the centerline of the keel to some point on either side of the transom; this will show you whether transom is true with the centerline of the keel. When you have it in the right position, fasten well with staylaths to prevent it from getting out of shape.

Part II

DIMENSIONS.

Length, over all	25 feet,	9½ inches.
Length, waterline	24 feet,	0 inches.
Breadth, extreme	5 feet,	0 inches.
Breadth, waterline	4 feet,	5½ inches.
Draft to Rabbet	0 feet,	11¾ inches.

YOU can then proceed to set up the molds, starting amidship or at either end. Set all molds forward of station No. 12, so that the forward edges are on the marks representing the stations, and those aft of station No. 12 so that the after edges are on the lines representing the stations. The reason for this is occasioned by the fact that the boat diminishes in breadth fore and aft of the station named and the edges of the molds being square would not admit of the shape of the hull being the same as the design. If you placed the center of the molds over the line representing the stations it would carry the forward edge of the mold forward of the stations from the stem to the amidship station, and vice versa from that point aft, which would increase the size of the sections. You could bevel all the molds if you desired, but in order to do this you would have to go to a lot of unnecessary labor.

Fasten the molds to the keel with screws, plumb the face of mold and also plumb the centerline, and when you are sure that it is in the correct position fasten with staylaths to the rafters. Set up all the molds in this manner, and before you put on any battens go carefully over each mold to avoid any possible mistake. Stretch the chalk line from end to end in a line with the L.W.L., then go carefully over each mold to ascertain that the L.W.L. marked on same is not above or below the chalk line.

If correct, you can proceed to put on the battens. These can be made of either spruce or yellow pine. If the latter, make them about 1⅝ inches square, and long enough to reach from the stem to the transom; if of spruce, about 1¾ inches square. Get clear stock so that they will bend fair. Start with the upper batten, which you should place so that the lower edge touches the marks representing the sheerline, fasten the forward end in the rabbet of the stem with a screw, bend carefully around the molds and fasten to each one until you get to the last two molds, and before fastening to them cut the end of the batten so that it will fit the rabbet of the transom; after fastening this batten you should put in the upper batten on the opposite side. The battens should be spaced about 9 inches apart at the largest section or station, No. 12, and about equidistant at the other molds, etc.

If, by any chance, you find that the rabbet on the stem or transom is not cut to the correct bevel you should trim it fair before you fasten the batten. I should suggest that you put two battens close together at a point above six inches out from the keel on either side, or else use a batten a trifle heavier than the others on account of the extra strain in making the short bend in the frames after they have crossed the keel. You will save considerable time when you start to bend the frames into the battens if you will mark on the keel, the upper batten, and one of the battens near the turn of the bilge, the position of the frames, as it will save you making a number of measurements.

The specifications call for frames ⅝ x ⅝ inches, spaced 6 inches center to center, and to be continuous from sheer to sheer from station No. 6 aft. You will need some good clear white oak. The butts of young trees furnish the best stock and they should be free from knots and straight grained.

Unless you have a steam box you will have to make one to steam the frames in before you can bend them in the boat. I know of instances where the frames have been bent to the required shape after being immersed in a trough of hot water, but the making of a steam box is a simple matter and as the longest frame will not be more than 9 feet long, allowing for a surplus on either side, you can get along very well with a box about 12 to 14 inches square and 10 feet long. Use spruce or pine, and make the joints tight, either by putting in strips of flannel laid in white lead or calk the seams with cotton. The latter method is preferable. Close up one end, and make a door for the other end. It is quite immaterial how you procure the steam for the box. You can use an old iron kettle or wash-boiler, or anything of that description that will be suitable for boiling a quantity of water in. Pipe from the kettle or boiler to the under side of the steam box. The steam does not want to be dry and hot, but should be wet, and in order to obtain the best results you will not need to have a great amount of heat under the boiler, simply enough to keep generating steam. Put a number of frames in the box and steam for 10 minutes; take out one of the frames and see if it bends easily; if so, you can start to bend in the frames, otherwise you will have to give them more time to get thoroughly saturated with the steam.

You will find that the bending in of the frames is a mighty slow job unless you can get some one to help you in handling them. Take one of the frames and start to bend it inside of the battens. You will find that it readily takes the required shape, and if you have someone to help you he can nail the frames to the inside of the battens as you bend them in. The frame should be sprung over the keel and first fastened to the strong or double batten next to the keel, and then work outward and up. Use small common wire nails to fasten the frames to the battens, as they will have to be taken out when you remove the battens. The frames, where they cross the keel, should be fastened with galvanized iron nails in preference to copper, as they are much stronger. If you prefer to use copper, get the hard nails for this part of the work, for while they will easily drive through the frame while it is wet and pliable, they will not drive easily into the oak keel.

I should suggest that you start putting in the frames at the largest sections, and by doing so you will have acquired the knack of bending them, and gained some experience before you start the bending of the frames at the after end where the topsides "tumble home." Frames forward of Station No. 6 do not cross the keel, and the heels of the same will have to be cut to fit properly, and the same thing applies to the frames in the wake of the stem and knee. The heels of these frames must be well fastened to the keel or stem.

When you have finished putting in all the frames, you can get out the sheerstrakes. If you are going to finish the sheerstrakes in natural wood varnished, you had better use either oak or mahogany. The latter material is called for in the specifications. The thickness is to be 7-16 inch. Procure a board long enough to extend from the stem to the transom in one piece, if possible, and if you intend to use mahogany there should be no difficulty in obtaining boards of the requisite length.

In order to get the shape of the sheerstrake you will need a staff or spiling batten, and this can be made from a piece of pine, cedar or spruce about ¼ inch thick, 8 inches wide and long enough to extend from end to end, or you can fasten two pieces together to make the required length. Take this spiling batten and bend around the outside of the frames below the line representing the sheer. If the second batten from the top is in the way, it can be removed without fear of disturbing the frames. See that the spiling batten lies flat and close to the frames, and don't try to spring it edgewise. Fasten it temporarily in position with clamps or small brads, then mark on the battens at each mold a line corresponding to the stations. Take your compasses, set them open a trifle wider than the greatest distance between the edge of the spiling batten and the sheerline on the molds and at each of these places sweep in a segment of a circle on the batten, cutting the vertical lines corresponding to the stations

Remove the spiling batten, and using same as a guide you can proceed to mark out the sheerstrake by proceeding as follows: Place the spiling batten on the piece of wood you intend to use for the sheerstrake, leaving the upper edge far enough away from the edge of the board to make sure that you will have room enough to get the proper curve without going beyond the edge of the board; take your compasses and set them open to about the same distance as you used when marking the arcs or segments of a circle on the spiling batten, set the spur leg on the points where the arc crosses the straight line and draw a segment of a circle on the board. Repeat this at each place you have marked, and then cut the arcs on the board with a line carried through the straight line on the spiling batten. Remove the spiling batten and take one of the battens you used in laying down the lines, bend to the points where the right lines cross the arcs and when you have it so that it cuts all the points draw a line through them, cut the board to this line and you have the shape of the upper edge of the sheerstrake. The shape of the lower edge is easily obtained, but you must first determine the width you desire to have this sheerstrake. It should not be more than 4½ inches at the widest place, which would be between stations No. 12 and No. 14. I would suggest that you make it 4 inches wide at Station No. 12, 3 inches at Station No. 2 and 3¼ inches at Station No. 23. Mark these widths on the board you have cut for the sheerstrake, measuring from the finished edge, and then, with a strong batten bent to the required shape, draw a line cutting the points you have marked, which will produce a fair curve, saw and plane to line and you have the shape of the sheerstrake.

You will need a duplicate of this for the opposite side and you can lay same on the board that you intend to use for the other side, mark around it, and cut to shape. Take one of these strakes and proceed to set in position preparatory to fastening, bring the upper edge to within half an inch of the sheerline, for, as before mentioned, this line represents the top of the planksheer, which is to be half an inch in thickness; use your clamps to hold it in place, fit the forward end in the rabbet of stem and fasten with brass screws

The fastenings in the frames are to be copper nails and the round wire nail would be the best. Sink the heads of the fastening into the planking deep enough to allow for covering them with wood plugs. Put one fastening through each frame at the upper and lower edges alternately; there are more fastenings to go through the sheerstrake and frames when the clamps and sill are fitted in.

When you have fastened the sheerstrake to within about 3 feet of the transom, you should cut the end of the plank so that it will fit nicely into the rabbet. Take your time in doing this, for if you cut it too short you will spoil the sheerstrake. When fitted, fasten with brass screws and finish fastening the other frames. Repeat this operation with the sheerstrake on the opposite side, and after putting on the burrs and riveting the fastenings you can give your attention to the garboard, such being the name of the plank next to the keel.

To obtain the shape of the garboard you will have to proceed in much the same manner as you did when get ting out the shape of the sheerstrake. In this case, however, you will work from the rabbet in keel instead of the marks on the upper part of the molds. A great deal more care is required to fit a garboard properly than a sheerstrake, as it is absolutely necessary that it should fit the rabbet in the keel very closely. Take your spiling batten, or if you have some more material of about the same thickness use that in preference to the spiling batten, for if you trim same to fit the rabbet as I am going to suggest, you are apt to spoil the batten. Take the spiling batten or a piece of wood of about the same dimensions and bend to the frames close to the keel; put on this batten some marks corresponding to the molds or other marks that you put on the keel, taking due care that the marks are at right angles to the keel, then take your compass, open them a trifle wider than the greatest distance from the rabbet to the edge of the batten, and, keeping the spur leg in the corner of the rabbet, draw a series of arcs across the right lines you have drawn on the batten. When you work toward the forward end where the rabbet on the keel connects with the rabbet on the stem, you will need to make the right lines close together in order to get the correct shape at this sharp turn. When you have got all the points properly marked remove the batten, and then, after drawing a line through the points, cut it to the required shape and fit it to the rabbet. If there are any places that need trimming, you can readily see where to pare, and unless you have made a poor job of taking the spiling or else not trimmed the batten carefully, you will have but very little work to make the batten fit nicely into the rabbet. When properly fitted, you can mark out the shape of the lower edge of the garboard on the piece of plank you intend to use for this purpose. The shape of the other edge is obtained in practically the same manner as you did the lower edge of the sheerstrake.

As each builder has his own ideas regarding the width of the garboard, either amidship or at the after end, and how far up on the stem he will have the forward end, it is scarcely worth while trying to set down any rule for this. You will not be very far astray if you make the garboard about 6 inches wide at Station No. 6, 5 inches at Station No. 2, and 5 inches at the transom. These dimensions need not be followed exactly, but should be kept in about the same ratio.

I will assume that you have learned, by this time, that the edge of the planks, whether garboard or any other strake, should be a fair curve. When you have one of the garboards cut to shape and planed fair and true on the edges, you can draw a duplicate of same on a piece of board for the garboard on the opposite side, and then you can proceed to fasten in position the one you have already fitted However, before you do so, it is necessary to put stopwaters in the joints where the stem connects with the keel. Bore a hole ⅜ inch diameter through the joint in the rabbet, and fill same with a soft pine plug. This will prevent the water working through the joint when the garboard seam is calked; also put a stopwater through the joint at the after end of the keel and transom knee. You will probably find it necessary to steam the forward end of the garboard, as there is considerable twist to it. There is no need to steam more than 5 or 6 feet of the forward end.

When you have it well saturated fit the forward end in the rabbet, taking special care to have it come in the same position as it was when you took the spiling and fitted it. Fasten with brass screws, spacing them about 3½ inches apart until well around the curve at the forward end. Fasten the upper edge of the plank to the frames with copper nails, first countersinking for the head. The fastenings in the lower edge are to go through the keel above the rabbet, one fastening to go through the keel and each frame, and one through the keel between each of the frames. When boring for the fasaenings, take care to have the hole through the center of the frame. The frames that cross the keel will not touch the center of the garboard, it simply bears in the rabbet and on the outer edge, and inasmuch as there should be one fastening at each frame through the center of the garboards if they are made the width suggested, or wider, and it will be necessary to cut some wedge-shaped pieces of wood to fit between the frames and plank at these points before you rivet them, otherwise when you riveted the nails you would draw the center of the plank out of shape. The garboard on the opposite side can be fastened in position, and you should bear in mind that the upper edge of the forward end ought to be in a direct line across the stem from the one on the opposite side.

Before getting out or putting on any more of the lower planks, it would be well to fit the keelsons and fasten them in position, as you can do this so much easier at this time than after the planking is in position; but in order to set the keelsons it will be necessary to remove the molds, and before you do that you had better put on three or four strakes on either side below the sheerstrake. so as to bind the frame together, and make sure that it will not change its shape when the molds are removed.

In order to find out the widths of the strakes you must first determine how many planks you will have on either side. Start with the frame that has the greatest distance between the edge of the garboard and the sheerstrake, and space out the planks on same. The widest planks should be nearest the keel and the narrowest at the turn of the bilge. If you made the garboard 6 inches wide at the frame you are measuring on, you could space off 5½ inches for the next plank, then 4½, 4¼, 4, and the balance about 3½ inches. Mark on the stem, around the transom, and two intervening points the spacing of the plank, keeping the widths in about the same ratio. When you have them marked on the different frames you will be able to ascertain how the plank will appear when in position, and if these points do not seem to come in a fair line go carefully over the spacing again, find out where the error is and rectify it.

All the planking of the boat should be widest at the 'midship section, and from this point forward and aft they should have a gradual taper, and not be wider either forward or aft of the 'midship section. This does not apply to the garboard, for in some instances the shape of the garboard is very peculiar, and if the boat had a different type of stem it would not apply to the planking.

You will not need any instructions how to get out the planks, for the marks on the frames will give you the widths at the different points, and you will work first from the lower edge of the sheerstrake, taking the spiling from same, transferring it to the plank, cut to shape, then set off the widths at the proper places, draw a line through the points and cut the lower edge to shape. When you have the plank cut to the required shape you will find that if the edge of the plank is left square with the face that when you set this edge up to the edge of the sheerstrake the seam is open wider in the center than at the ends, and you must trim the edge so as to have the seams open on the outside about 1-16 of an inch the entire length. Take the bevels off the edge of the sheerstrake at intervals and plane the edge or the plank that is to be fastened next to same to correspond to these bevels or nearly. so, as the joints of all the planks should be tight on the inside and open on the outside so as to receive the calking cotton. You must keep in mind the fact that all the planks will have to be bevelled. Another thing to which I will call your attention at this time is the fact that, after you have put on one or two strakes below the sheerstrake you will have to cope or hollow the planks on the inside so that they will fit close to the frames, and as the specifications call for the planking to finish 7-16 inch in thickness, the planks that are coped will have to be thicker than the others. If the coping amounts to about 1-16 of an inch or more, have the plank that much thicker and when you have it coped to fit set your gauge to 7-16 inch and from the inside of plank mark both the upper and lower edges, and then bevel them from the outside of the plank to the gauge mark, the bevel being about 1 inch in length. This will materially assist you when you plane the outside of the planking, for if you jack the plank down until you are close to the seams that are bevelled in this manner you can be reasonably sure that the plank will be an even thickness.

After you have fastened four strakes on either side you can remove the molds. As you take each one out, fasten a brace across the upper edge from one side to the other, nailing them to the frames to prevent the boat spreading, and also secure with staylaths overhead at intervals.

You can now give your attention to the keelsons. These should be in one length. Select two nice, clear pieces of spruce and plane to the required thickness; measure out from the center of the keel to the inside of the keelsons. The plans show that these are to be spaced 17 inches apart, therefore, you will measure out from the centerline half that distance, and mark the frames at intervals. Take one of the pieces and set it to the marks on the inside of the frames. It will not touch any of the frames in the wake of Station No. 12 by several inches, but if you fasten it in some manner or another, keeping it upright, you can get the shape of the lower edge by taking your compasses and marking it at intervals in much the same manner as you take the spiling of the plank. Remove the piece and after bending a batten to the points and drawing a line through same, cut to shape. If you feel satisfied that you can get the correct shape the first time, it would pay you to take the trouble of marking both sides of this keelson and thus get the bevel, and also mark the places that have to be cut out to allow it to fit over the frames. If you are not sure of your ability in this line, you had better trim roughly as suggested, then set it in position, fasten with clamps or some simple method, and proceed to mark on both sides the location of the frames, setting a straight line on either side of the frames, and the compasses will be brought into play again to mark the required depth that the keelson is to be cut to allow it to fit over the frames.

I would suggest that you allow a depth of about 1-16 inch more than the thickness of the frames, so that the lower edge of the keelson will project below the frames, and then when it is in position you can plane off the surplus stock; and if this is done carefully the inside of the plank will fit tight against the lower edge of the keelson. Both keelsons will have to be fitted in the same manner and the height at the different stations can be taken from the plans. When you fasten them in position, use either galvanized iron nails or brass screws, bore through the frame and fasten into the keelson at each alternate frame.

As the bilge clamps are easier to set in position before the boat is planked, you might as well put them in before proceeding with the planking.

The specifications call for the bilge clamps to be made of spruce ⅞ x 3 inches. Select good, clear stock and work to the required shape. They are to be tapered to 2 inches at the ends, and this taper should be about 8 feet long at the forward end and about 6 feet at the after end. Plane and finish them before you fasten in position. Spring the center in to the frames at the same point as shown on the plans, and the forward end to approximately the same point as shown on plans, although you must be guided to a great extent by the shape it assumes. It is not advisable to spring it edgewise, and it should be allowed to assume such a shape that it will lie close to the frames and not have either of the edges standing up from same. When you are sure you have it in the correct position, fasten with copper nails 3-32-inch diameter and rivet over burrs on the inside. The fastenings should be alternately near the upper and lower edge, one fastening through each frame. When the bilge clamps are in position, you can start planking again. Start with the plank next to the garboard and work upwards. After fastening a plank on one side fasten the duplicate on the other side; don't put two or three planks on one side of the boat and then two or three on the opposite side. Watch the forward and after ends of the planks as you progress and see that you do not gain faster on one side than the other, for if there is anything that looks bad it is to have the wood ends on one side of the stem or transom higher than they are on the other side. However, if you are careful to shape and fit the planks to the marks you have made on the frames you are not likely to experience this trouble.

The last plank to be fitted in is called the shutter, and I should advise you taking a spiling for both edges of this and for both sides of the boat, for if there is any variation in the space on either side it will be necessary. Take particular care to have the shutter large enough; it should be a good driving fit, and it will then help to tighten up all the planking. You will find that you cannot rivet the nails over burrs where the fastenings come in the wake of the bilge clamps, and the best plan would be to fasten the planks at such places with brass screws.

If, for any reason, you prefer to use planks that are not full length you can do so. The joints should come between the frames and be well fastened to oak butt blocks. Make the blocks the same thickness as the frames and cut to the shape of the inside of the plank. The blocks should extend from frame to frame and be about ½ inch wider than the plank which butts on them.

The seams should be roughed down with a jack-plane before the boat is calked, but there is no need to try to plane the outside smooth until after the calking has been done, for then it can be planed to much better advantage.

If you have decided to calk the boat yourself, you can now proceed with same; but I should strongly advise your procuring the services of an experienced man. In spite of the fact that it looks easy to do such work when you watch a calker driving in cotton, there is a great deal more to it than appears to the amateur. It is not simply a matter of driving in cotton; a great deal depends upon how much cotton is required, and how hard to drive it. If you want to finish the job yourself, and would like to do this calking, take a small strip of fine calking cotton and drive it into the seams, using a thin calking iron and mallet. This cotton should not be driven in the seam in a line like a piece of rope, but keep catching back a loop every inch or so, and tap it in place until you have proceeded in this manner for a few feet, then go over it with the calking iron and drive it in until it is about ⅛ of an inch below the surface of the plank. If you find it drives in easy use more cotton, if too hard, less cotton. The man who makes a business of calking when he finishes to a certain point leaves the end of the cotton showing and does not drive this in place until he starts to continue the calking of the seam, and in this manner avoids missing any places. You will have to calk the seam in the stem, and also around the transom. The garboard seam will probably require more cotton than any of the others, but this depends entirely upon how well you have done your work, whether the seams are large or small.

The deck clamps, which are to be of spruce, ⅞ x 3 inches amidship, tapered at the ends the same as the bilge clamps, can now be gotten out. Select good clear stock; small fine knots are not detrimental, but avoid any stock with large knots or short grain, as the clamps should be strong. Set the upper edge of the clamp the proper distance (see plans) and fasten in position, using bolts or nails 3-32-inch diameter. The fastenings are to be through sheerstrake, frame and clamp and to be riveted over burrs on the inside. Do not place them in a direct line; they should be staggered, being placed alternately near the upper and lower edges of the clamp. When you have finished fastening the clamps you can get out the quarter knees, fit them and fasten to the clamp and transom, and then get out the breast hook; fit this carefully to the stem and inside of the sheerstrake and fasten securely. It is a good plan to make the breast hook high enough in the center to allow of it being worked down to the crown of the underside of the deck.

The shelves are to be of spruce, to finish 7/8 x 2 5/8 inches, and you will require two pieces long enough to make them without any joint; cut and plane to the proper size, and then bevel the edge so as to have the upper side on the same bevel as the planksheer, clamp in position and fasten through the sheerstrake. It is advisable to fasten this shelf through each third frame, selecting such frames as have the fastenings through the clamp near the lower edges, otherwise you would in all probability get too many fastenings through the heads of the frames, and thus tend to weaken them. Put in a few fastenings through the shelf into the upper edge of the clamp; one about every 20 inches would be sufficient.

For the after deck you will require three beams 7/8 x 1 3/4 inches. The camber or crown of the after-deck beams is 2 1/2 inches in 4 feet. Cut the ends of the beams into the clamp about 1/4 inch and fasten to same. Set in between the beams a block of oak as shown on the construction plan for the tow-post to be mortised through. Trim the quarter knees to the same crown as the deck beams and the after deck frame will be ready for the deck plank.

The beams for the forward deck are segments of circles, each one having a different radius. The simplest way for you to get out these beams is to saw them to shape, for if they were bent to shape you would have to make a separate mold for each one. One method of getting the curve of these beams is to sweep in the curve by using a long staff with a pencil attachment at one end, but as this requires a long arm to get the proper radius the best and simplest way is to construct the curve geometrically. There are several methods of constructing a segment of a circle, that has a radius greater than can be drawn with the usual compasses, but I will deal with one of the simplest for the benefit of those who are not well versed in geometry.

We will take as an example the beam at the forward end of the coaming; the center of the beam is 3 3/4 inches above the edge of the sheerstrake, and the width of the boat at this point measured inside of the upper edge of the sheerstrake is 45 inches. Using these dimensions for the base and height, you can proceed to construct a curve as per Fig. 10. Draw a straight line, A B, equal to half the length of the beam, from the point A draw A C perpendicular to A B and in length equal to the crown of the beam, draw a line from B to C, and from the point B draw B P perpendicular to B C; through C draw C P parallel to A B. Divide A B and C P into any number of equal parts, the same number in each, and join those opposite to each other as D G, E H, F I. Through B draw B O perpendicular to A B, cutting C P at O. Divide B O into the same number of equal parts as A B or C P and join C with each of the divisions. Then through the point C, the point a where D G cuts C J, the point b where E H cuts C K, the point c where F I cuts C L, and the point B, draw the curve C a b c B. This curve will be a segment of a circle, but as it only represents half of the beam, and if A N is made equal to A B and a similar construction made, then B C M will represent the curve of the beam. When you have constructed one of these arcs you will find that it is a very simple matter to make the others, and if you set out each beam in this manner you will find that the centerline of the forward deck is a straight line, and the curve of the deck fair and true. When you have cut out the beams, fasten them to the clamps and then get out the coaming knees. Make these of spruce, fasten well to the beams, etc., and trim the upper side so that you will have a fair curve. Trim the inside edge of the knees so that they are plumb and cut to the same curve as the forward end of the coaming. Fit in between the beams blocks for the cleat and cowl ventilator as per plans.

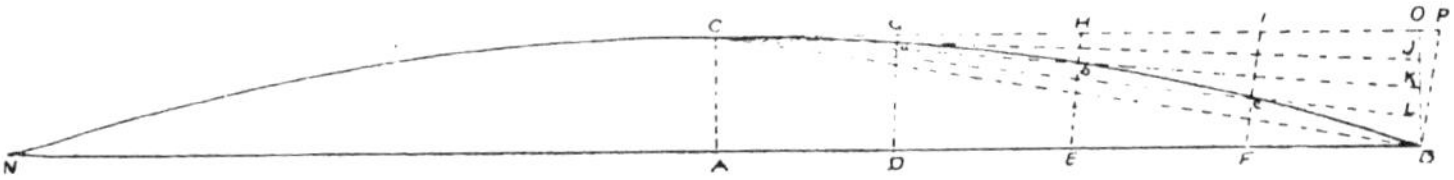

Fig. 10—Showing Method of Constructing Curve for Beams

You can now give your attention to the pieces to make the planksheer, which is the plank on the outer edge of the deck. The specifications call for this to be made of mahogany 1/2 inch thick; it is to be 4 inches in width, from the after end of the cockpit to a point about 2 feet aft of the forward end of the coaming around the hood, and from this point forward it is to have a slight taper; the after end should also be slightly tapered. You will have to cut this plank to the proper shape before starting to fasten it, and if you have been able to get out the proper shape of the different planks there is scarcely any need for me to explain how to get the shape of the planksheer. You will not be able to get it out of one piece on account of the curve, and therefore I would suggest that you have the joint about midway between the forward and after end of the engine hood. Naturally you will have to be guided to some extent by the length and width of the stock you can procure. When you have the pieces cut out for the planksheer, begin to fasten at the forward end and work aft; fasten in position the pieces across the after end of the after deck, and strike a centerline on the deck beams fore and aft. This line is to be used as a guide for laying the deck plank.

The specifications call for the deck plank to be made of mahogany 1/2 inch thick and 3 inches in width. Starting with the after deck, bring the edge of the first piece up to the centerline, fit the after end and fasten with brass screws, countersinking the heads so that the fastenings can be covered with wood plugs. Work in the plank on either side, fastening them as soon as you fit them, and before fitting the next one. Proceed in this manner until the deck is covered and you can proceed in the same manner to cover the forward deck. If you are particularly anxious to make a very nice job of the deck, you can taper each piece of plank, and not carry them on a straight line, as suggested. To obtain the shape of the planks if worked in this manner, you would have to decide first how many pieces of plank you intended to use; space them on the beam having the greatest length and then space the rest of the beams for the same number of plank. It is much the same operation as I suggested for setting out the planks after you had the sheerstrake and garboard in position

The under side of the plank for the forward deck will undoubtedly have to be coped to fit the round of the beams.

Calk the seams lightly with a fine thread of yacht cotton, and if the deck is to be finished bright, plane roughly to the required shape and fill the seams with white lead putty, unless you prefer to use the elastic seam composition or marine glue. The final finishing of the deck should not be considered at this stage of the game; it is far better to leave this until all the principal woodwork is finished.

The engine-bed should be fitted and fastened before you start to put in the beams for the cockpit deck.

The length of the bearers for the engine-bed will depend to a great extent upon the engine you intend installing, and that will also govern the thickness of same. The specifications call for the bearers to be sided 2 inches. They should be fitted over the frames the same as the lower edge of the keelsons, and be well fastened to same. The height of the upper edge of the fore and aft bearers would be determined by the distance the underside of the engine-bed where it rests on the bearers is above or below the center-line of the crank-shaft. You will have to obtain the measurement before fitting the fore and aft bearers, and should also obtain the width that will be required between the bearers.

To obtain the line of the center of the shaft, you should use a piece of very hard laid fish line, or, better still, a piece of piano wire. Set out the height on the post supporting the after end of the stocks, and another point inside of the boat well forward of where the flywheel of the engine will set. Fasten a piece of board to the frames and have the upper edge a trifle higher than the center of the line at this point, cut a nick in the edge, so that when the wire sets in same it will be the proper height, take the other end through the hole in the keel, and as the form will prevent your drawing the line directly aft to the point representing the center of the shaft, you will have to bore through the form or you can cut part of it away, so long as you leave enough to support the boat properly. Bore a hole through the center of the end-post and, pulling the wire through this to the required height, draw it as taut as possible, and you can then measure to this line, either up or down as the case may be for the upper side of the engine bearers.

When you have them fitted and fastened, if there is room enough between the after side of the flywheel and the forward end of the crank-case to admit of using a good-sized athwartship bearer, it is a good plan to fit one in position and if possible to bolt through the keelsons, fore and aft bearers and the athwartship bearer, as it makes the engine-bed more solid and tends to eliminate the vibration of the engine by distributing it more evenly. One of these athwartship bearers at the after end of the engine is also to be recommended.

If you have your engine on hand, it would be advisable to set it in position and fasten to the bed and then line up the shaft and fit the shaft tube, otherwise you will have to leave part of the cockpit deck open until the shaft tube is fitted. The drawing, Fig. 14, shows the type of shaft tube such as I use on my boats, but if you prefer you can purchase one of the adjustable shaft tubes that are now on the market. You can put in the exhaust pipe and water connections and, in fact, finish all the piping of the engine much more readily at this period than later on, when you would have to take up floors, etc.

Bore the hole through the transom knee and fit in the rudder port, which is to be made of brass pipe fitted with a stuffing-box and lock-nut as per Fig. 20. The lower end of the port should be threaded and extend through to the under side of the keel, and fit tightly in the hole to obviate any chance of a leak around same.

The cockpit beams should be placed in position and fastened. They are to be of spruce, sided ⅞ inch, molded 1¾ inches. Space them as shown on the construction plan, fastening the ends securely and setting short, upright pieces for stanchions over the keel wherever necessary. The cockpit deck is to be 1 inch lower at the after end than at the forward end, in order to drain off the water. Strike a line through the center of the beams, and fasten the deck plank to the beams. The deck planks are to be of spruce or pine ½ inch thick, and whether you follow the specifications as to width or not is immaterial, but it is not advisable to use very wide boards on account of the

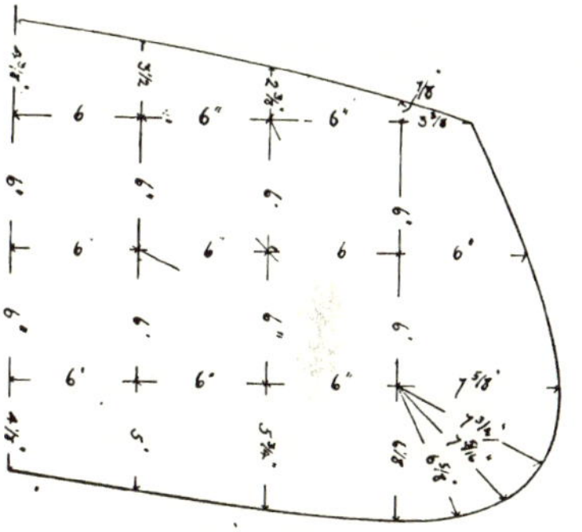

Fig. 13—Transom

tendency to shrink and swell, and the greater the number of seams, the less the shrinkage and swelling will be perceptible.

The coaming should be made of mahogany or some hard wood. Select nice straight-grained stock; it is to be a half-inch in thickness. If you have a good wide board long enough to work from a point aft of the engine hood, clear around the forward end to the opposite side, it would be advisable to use it, otherwise you can make the joint at the forward end and secure it with a butt block on the inside. You will have to use your spiling batten or some thin stock to get the correct shape. The width you can take from the construction plan; steam the forward end, and bend over a form of the required shape. Fasten well with brass screws from the inside to the shelf and coaming knees. It would be a good plan to leave the coaming a trifle higher than shown on the plan and to trim this down to the proper height after it is fastened in position. A surplus of about ¼ inch would be ample. Where the butts occur on either side of the cockpit you need not use a butt block on the inside unless you prefer same. They are rather unsightly at the best. If you intend to carry an oar on this boat you could put a butt block on the outside of the coaming and fasten the oar-lock sockets to same, but the best method is to make a tight butt joint and then bore a hole ¼ to 5-16-inch diameter down through the center of the joint and drive in a dowel. The after ends of the coaming where they set on the deck can be fastened from the under side or from the upper side if preferred, and the heads of the fastenings covered with wood plugs. Round slightly the upper edge and shape the after ends as per plans. A light thread of cotton should be worked in around between the coaming and the edge of the deck.

The bulkheads at the forward and after end of the cockpit should next claim your attention. Build up a suitable frame on the deck and around the inside of the plank and cover with tongued and grooved mahogany ⅜ inch thick and about 3 inches wide. It is preferable to use bevel-edge staving, to that with a bead, as it is so much easier to keep

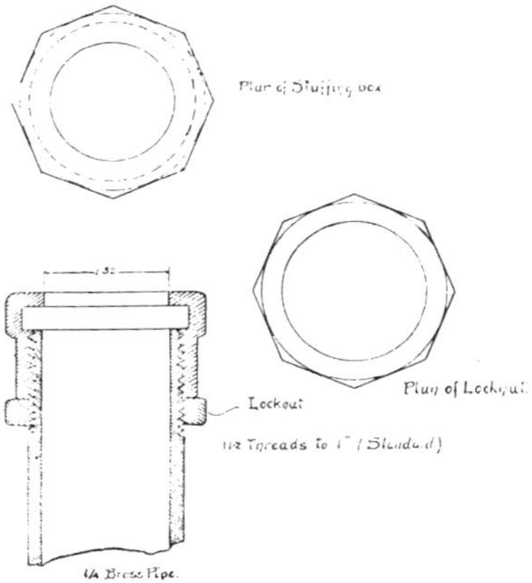

Figure 20—Rudder Port

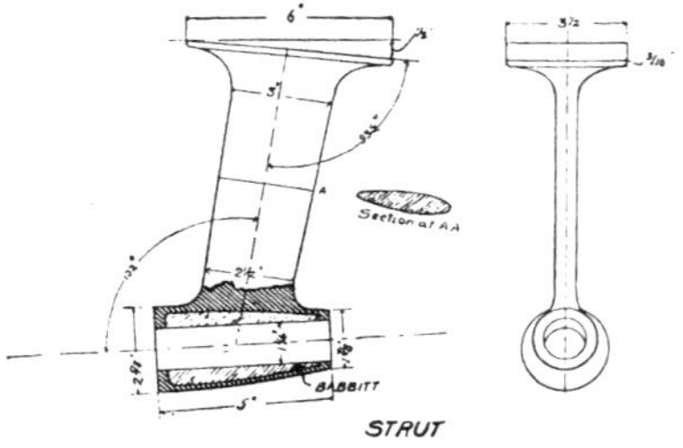

Figure 17—Strut

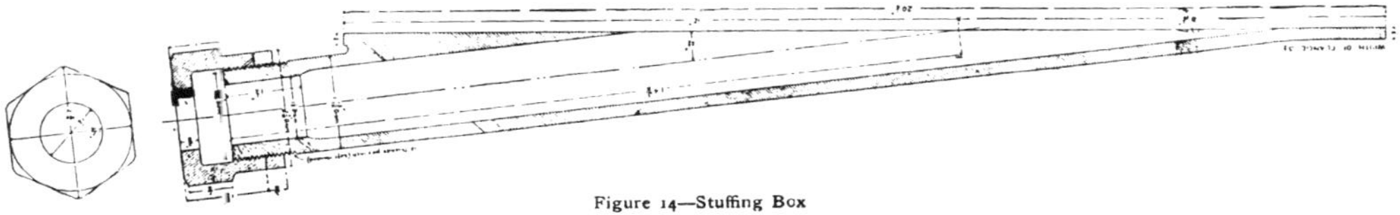

Figure 14—Stuffing Box

clean and to scrape, when the latter becomes necessary. Fasten the lower ends of the staving to the frame on the deck, and to the upper frame; make sure that you keep the line of the staving vertical and above all have good joints, as such a bulkhead materially strengthens and stiffens the boat if properly fitted.

Where any fittings, such as auxiliary shaft for starting engine, controls, steerer, etc., are cut through or fastened to the bulkhead it would be well to reinforce same on the inside with oak blocks ½ inch thick well fastened to the staving. The bulkhead at the after end should be staved up in the same manner as the forward one, and as you may require to get at the steering apparatus it would be well to cut a door in same, the lower edge to be not less than 2 inches above the deck; if it were flush you would find that considerable of the rainwater, and such water as you might ship, would run down into the bilge instead of overboard through the scuppers.

These scuppers should be made of 1-inch inside diameter lead-pipe; fairly heavy stock, weighing about 4 pounds to the foot should be used; set them in the after corners of the cockpit deck, recessing the wood so that they can be flanged over and still be flush or below the level of the deck. The holes in the underbody where the scuppers pass through should not be directly below those in the deck, but lead aft two or three inches and the lower ends should also be flanged and the wood recessed to receive the flange so as to insure a good, smooth job. Fasten around the flange at both ends with fine brass screws or copper tacks.

The ceiling is to be of mahogany, 5-16 inch thick, 2 inches wide, and should be made of tongued and grooved stock with the edges bevelled, to match the bulkhead staving. Fasten well to the frames, the ends to butt against the staving of the bulkheads. The lower edge where it meets the deck should either be set into a rabbeted piece of spruce that was fastened closely to the deck in order to insure a tight joint, or you can take a piece of canvas about 5 to 6 inches wide, set the lower edge in white lead and fasten to the deck with copper tacks, then fasten the upper edge to the frames and apply a coat of heavy white lead paint and set the lower piece of staving on the edge of the canvas and fasten at every frame.

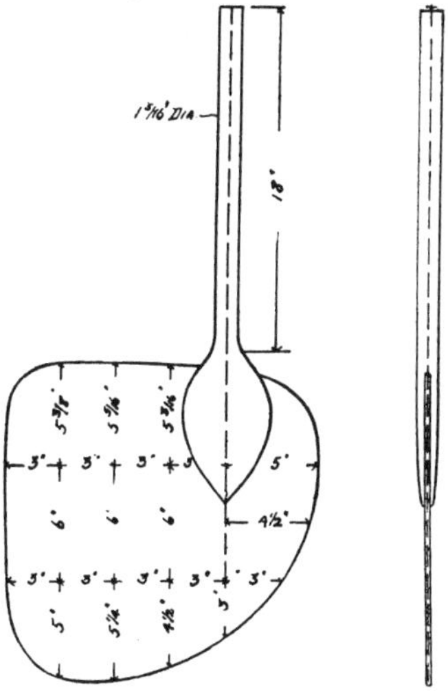

Figure 18—Balanced Rudder

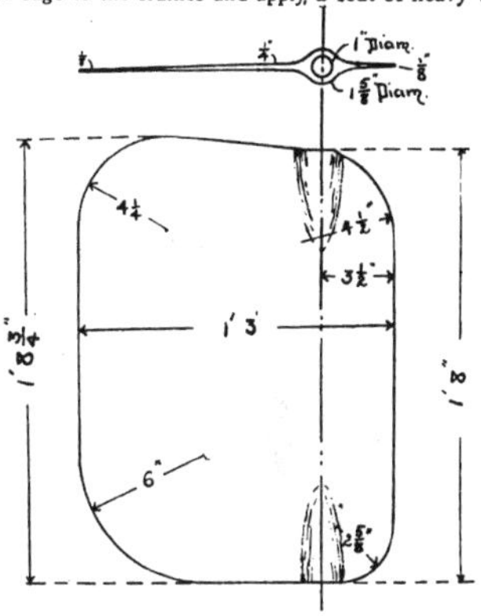

Dimensions of Rudder Blade

Figure 19—Skag Rudder

Frame up the helmsman's seat with spruce and cover the forward and after sides with staving to match the bulkhead. Cut a good-sized limbur through the staving and frame on each side so as to allow any water that may come in forward of the seat to run aft, and out through the scuppers. The gasolene tank is to be of copper, and is to be fitted under the seat, and when in place you can put on the top of the seat. It is advisable to put this on in such a manner that it can easily be removed, if at any time it is necessary to take out the tank. Cut in flush with the top of seat a large deck-plate directly over the filler plate of the tank.

The seat and backboard at the after end are so plainly shown that there is no need to do anything more than mention that they are to be fitted and fastened. The thickness is given in the specifications and the other particulars can be taken from the plan.

The next most important thing is the hatches, and frame of the hood covering the engine. For the strongback, you will require a piece of spruce 2 x 8 inches and 5½ feet long. Shape this as per Fig. 21, with a waterway on either side, and hollow the under side to keep down the weight. Fasten this at either end with a small metal knee, so that if you want to take it out at any time you can do so and not destroy the woodwork. Make a form to bend the frames for the edges of the hatches and make them of oak; after bending let them set properly before you try to fit them; and another thing that may assist you would be to bend them with a trifle shorter turn than you really need. You can easily straighten them a little, but it is hard to get them to bend any more once they are set.

Make up the rest of the frame so that it fits easily inside of the coaming, and then cut out the beams, which are to be of spruce. Set the ends into the frame and fasten securely. When you have finished both frames, set a ledge around inside of the coaming for them to rest on; place the frames in position and ascertain that the upper side is fair; if not, plane it until it is fair and true and then you can start to cover the frame with the plank, which is to be of mahogany 7-16 inch thick.

The piece that fits over the strong-back is to be 6 inches wide, and the widths of the plank on the hatches can be determined by the stock you have on hand, but don't make them too wide. Either calk the seams lightly or batten them on the under side. Carefully smooth the top of the hatches and trim the outer edges fair with the coaming. The edges are to be covered with a strip of brass No. 14 gauge, 1 inch wide. Cut this to the proper shape to bend around the edge and fasten with countersunk-head brass screws—the edges of the brass should be carefully rounded. Fasten the hatches with brass or bronze hinges, the joints to be directly over the joint of the mahogany centerplank and edge of hatch. Suitable quadrants, or something to take the place of same, should be

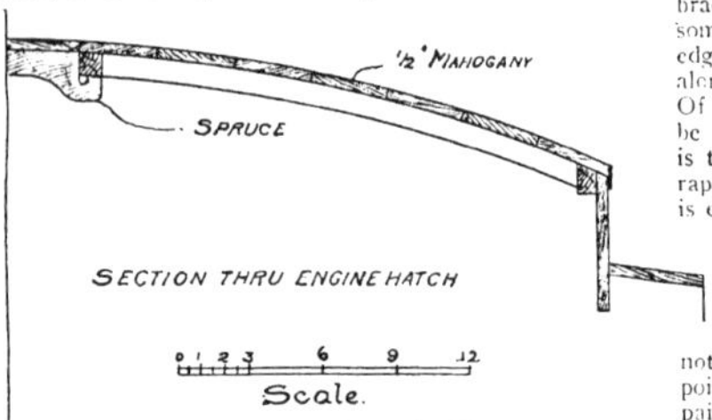

Figure 21—Hatch Over the Motor

fastened to the after end of the hatches to keep them open when desired.

After you have finished with the interior of the cockpit and coaming you can finish planing and smoothing the decks; set the bitt in the after deck, and you can turn your attention to the planing and smoothing of the planking. Finish it carefully, scrape and sandpaper, then fasten in position the pieces of half-round mahogany molding, and if you have decided to have a gilt stripe cut a cove in the plank at proper height, and then mark the waterline. I would suggest that you raise the line so that when you have the painted line it will show about 1 to 1¼ inches above the actual waterline.

There are a number of different ways to mark a waterline, and some of you may know a simpler method than the following, but this will give the desired results. Mark on the stem and the transom the height of the waterline, take two straight-edge planks, a little longer than the width of the boat, set the upper edge so that it is level with the point on the stem representing the waterline, fasten this piece of wood and make sure that the upper edge is level, using your spirit-level to make sure; repeat the operation with the other piece of board at the after end, then stretch a line from the upper edge of the forward cross-piece to the after one, draw the line very taut, take your spirit-level and bring it up to the line; see that the level only just touches the line and then where the upper edge of the level touches the hull make a point; repeat this operation at intervals all around the hull, then take a thin batten, something about ⅛ x 2 inches, and bend it to the point you have made, keeping the side of the batten vertical; fasten at intervals with fine wire brads, and when you have it fastened, take a scriber or some pointed instrument and scratch in a line along the edge of the batten. When the batten is removed sight along the line you have made, and it should be straight. Of course, it follows the shape of the boat, but it should be level. The idea in scribing or scratching in this line is that it does not get obliterated with painting, etc., so rapidly as if it was simply drawn on with a pencil, and it is easier to draw to a painted line than to a pencil mark.

There are still a number of things for you to do before the hull is completed. The strut is to be made and fitted, the steerer and steering ropes, the rudder, the deck fittings, etc., to say nothing of finishing the installation of the engine; but I do not see that there is any need for me to dwell upon these points any more than there is to try to explain how to paint the boat.

Unless you have good reasons for doing otherwise, I would recommend that you finish the boat entirely while she is on the stocks and if you have got to take her any great distance, build a cradle under her. This you can readily do. Then block up the cradle, remove the stocks and lower the cradle down to the floor or else to the height of the team if she has to be carted any distance.

When purchasing the paint or varnish, get the best; it will pay in the end. Don't hesitate to spend time finishing the boat properly, sandpapering down each coat before the next is applied.

Naturally there are many points I have not touched upon, but I believe that all the main points have been covered, and if I have made any grave errors I trust that you will overlook them.

Specifications for the Construction of a 25-Foot Runabout

BY FREDERIC S. NOCK

MATERIAL AND WORKMANSHIP: In carrying out these specifications, there are to be used only the best of materials and workmanship. All wood shall be sound and clear, all pieces to be cut fair with the grain and selected to have the grain follow the shape as closely as possible.

KEEL: To be of native white oak, molded 1¼ inches, shaped as per plans, and to be rabbeted to receive the garboard. The upper side is to be rounded as per detail plan.

STEM: To be a natural crook of oak or hackmatack, sided 2½ inches, molded as per plans. To be connected to the keel with a lock scarf and fastened with 5/16-inch diameter bolts riveted over washers. Heads of fastenings to be countersunk and covered with wood plugs.

TRANSOM: To be of white pine or mahogany, ⅝-inch thick, connected to the keel with an oak or hackmatack knee, sided 3 inches and molded as per plans. To be rabbeted to receive the transom plank, and to be fastened with 51 16-inch diameter bolts riveted over washers, heads to be countersunk and covered with wood plugs. Cheek pieces of oak ⅞-inch thick, 1½ inches wide are to be shaped and fastened to the transom to form a back rabbet for the plank which is to be well fastened with brass screws.

FRAMES: To be of young white oak, ⅝ x ⅝ inch, spaced 6 inches center to center. To be steam bent to shape and all frames aft of Station No. 6 to extend from sheer to sheer in one piece and to have two fastenings in keel. Frames forward of Station No. 6 that do not cross the keel are to have the heels well fastened to the keel.

KEELSONS: To be of spruce sided 1 inch, molded as per plans. To be spaced 8½ inches from the center to the inside. They are to be cut over all frames, carefully fitted and well fastened.

DECK CLAMP: To be of spruce, ⅞ x 3 inches amidships, tapered at ends to ⅞ x 2 inches. To be well fastened to the frames, and at each alternate frame to have one fastening through frame and sheerstrake.

BILGE CLAMPS: To be of spruce, ⅞ x3 inches amidship, tapered at ends to ⅞ x 2 inches. To be well fastened to the frames.

SHEERSTRAKE: To be of mahogany, 7/16-inch thick, and about 4 inches wide amidship, tapered at ends. To be well fastened to the frames with copper nails riveted over burrs. The fastenings through the upper edge on each third frame are to go through sheer strake, frame and shelf for the length of the shelf.

PLANKING: To be of clear cedar or white pine to finish 7/16-inch thick, to be in long lengths, and where butts occur, same are to be reinforced with oak backing. Fastenings to be copper nails riveted over burrs. Heads countersunk and covered with wood plugs. Inside of plank to be properly coped to fit the frames, outside to be planed fair and true and finished smooth. Seams to be calked with yacht cotton, payed with white lead and finished flush with white-lead putty.

SHELF: To be of spruce, ⅞ x 2⅝ inches, to extend the full length of the cockpit and hatch coaming as per plans. To be well fastened to the clamp, and at each third frame to have one fastening through frame and sheerstrake.

DECK BEAMS: To be of spruce, sided ⅞ inches, molded 1¼ inches, sawed to the required camber and spaced as shown on plans. Ends to be well fastened to the clamp.

COAMING KNEES: To be of spruce, 1½ inches thick, shaped as per plans and well fastened to the shelf and beams.

PLANKSHEER: To be of mahogany, ½-inch thick, 4 inches wide in the wake of the coaming and tapered at ends. To be well fastened to the deck beams, etc., with brass screws, heads to be countersunk and covered with wood plugs.

DECK PLANK: To be of mahogany, ½-inch thick, laid in strips 3 inches wide and well fastened to the beams with brass screws, heads to be countersunk and covered with wood plugs. All seams to be calked with yacht cotton and filled with elastic seam cement or white-lead putty, as desired.

COAMING: To be of mahogany, ½-inch thick, steam bent to shape at forward end, and to be well fastened to the shelf and coaming knees.

COCKPIT DECK BEAMS: To be of spruce, sided ⅞ inches, molded 1¼ inches, spaced 12 inches. Ends of beams to be well fastened to the frames and clamp and supported where necessary on spruce stanchions ⅞ x 1½ inches, set over the center of keel.

COCKPIT DECK PLANK: To be of clear spruce or white pine ½-inch thick, laid in narrow strips, to be fastened with galvanized nails or brass screws. To be either tongued and grooved or calked seams as desired. Deck to be made water-tight and to drain aft through lead scuppers 1-inch inside diameter.

BULKHEADS: The bulkhead at the after end of the engine compartment is to be framed up and covered with mahogany strips ⅜-inch thick, 3-inch face tongued and grooved. Wherever the fittings pass through the bulkhead, same is to be reinforced with blocks on the inside. The bulkhead at the forward end of the afterdeck is to be framed up and covered with ⅜ x3-inch mahogany the same as the forward bulkhead, and a suitable trap is to be cut in same to admit of access to the steering gear, etc.

CEILING: The inside of the frames above the cockpit floor are to be covered with mahogany staving, 5/16-inch thick, 2-inch face, ceiling to extend from the upperside of the deck to the underside of the clamp and to be well fastened to the frames with small finishing nails.

STERN SEAT AND LAZY BACK: To be of mahogany ¾-inch thick, shaped and fitted as per plans.

HELMSMAN'S SEAT: To be framed up as per plans and staved on the forward and afterside with mahogany ⅜-inch thick, 3-inch face. A copper gasolene tank 12 inches deep, 15 inches wide and 42 inches in length fitted with suitable swash partitions, is to be carefully fitted under the seat. Bottom of tank is to be 1½ inches above the cockpit floor, and suitable limburs are to be cut in the frame and staving to admit of allowing the water to run aft under the tank. The top of seat is to be of mahogany, ¾-inch thick, and fastened in position with brass screws. A suitable deck-plate is to be let in flush with the top of the seat over the filler of tank.

BITT: To be of oak or locust, 3x3 inches, to be set through deck and blocking and well fastened on the underside by an oak or locust wedge.

ENGINE BED: To be arranged to meet the requirements of the engine. The fore and aft bearers to be of oak, sided 2 inches, and to extend well beyond the foundation of engine fore and aft. They are to be well fastened and bolted through keelsons. Suitable athwartship bearers to meet the requirements of the engine are to be provided, one between the flywheel and base of engine, the other one well aft, and if there is room to admit of using same, there is to be one bolt through keelsons, fore and aft, and athwartship bearers at either end, bolts to have nut and washers on the outside of keelsons.

FENDERS OR RUBSTREAK: To be of mahogany 1-inch, half round, well fastened to the sheerstrakes, etc.

HATCHES, ETC., OVER ENGINE SPACE: A strongback of spruce with suitable waterways on either side and worked out on the underside as per plans to be made and well fastened to the coaming and bulkhead. The hatch frame is to be made of oak, curved part steam bent to shape and carefully fitted. Finished size of section of frame ⅞x1¼ inches deep. The beams are to be of spruce sided ⅞ inches, molded 1 inch, sawed to crown. Ends of beams to be cut into the frame and well fastened. Hatches and strongback to be covered 7-16 inch mahogany, in narrow strips, seams to be battened on the underside or calked as desired. Fastenings to be brass screws, heads countersunk and covered with wood plugs. The edges of the hatch are to be covered with a brass strip, gauge No. 14 stock, 1 inch in width, fastened in position with countersunk head brass screws. Three hinges of bronze, 18 inches in length, 6 inches center to center of joints are to be used for the hatches, and there is to be a bronze quadrant or slide for either hatch that will admit of them being held open.

FITTINGS: To comprise:—One shaft tube of bronze constructed as per detail plans with stuffing-box and locknut, flange to be carefully fitted to the keel and well fastened with brass screws. Bronze rudder, with cast manganese bronze stock, and hard rolled Tobin bronze plate 3-16 inch thick. brass rudder-port with stuffing-box and locknut, galvanized iron quadrant or sliding tiller, galvanized steel or Phosphor bronze tiller rope ¼ inch diameter, galvanized iron or bronze sheave leads of not less than 3 inches diameter for tiller rope. One adjustable auto steerer with wood rim wheel 12 inches diameter, one 12-inch bronze cleat, one 4-inch cowl ventilator, two 6-inch brass combination chocks, two 5-inch brass quarter chocks, two bronze flush flagpole sockets, one 24-ounce copper tank of about 30 gallons' capacity with suitable swash partitions, bronze filler, vent plate and supply connection. One 6-inch diameter deck plate over stuffing-box, one suitable deck plate over tank filler, three bronze hinges for hatches, two quadrants or slides for hatches, brass stem band, two flagpoles, one yacht ensign, one private signal or yacht club pennant, one pair of side lights, and one bow light, either galvanized iron or brass as desired. One 25-pound folding anchor, 100 feet 9-16-inch diameter Manila warp, one 6-inch diameter ship's bell, two Manila fast lines 30 feet in length, two canvas fenders 4 inches diameter, 16 inches long, and one bronze head boat hook with 7-foot handle.

PAINTING, FINISHING, ETC.: The entire interior of the hull to receive one coat of priming paint before being ceiled. The topsides to be primed and to receive three coats of pure white lead paint, each coat to be well rubbed down before the second coat is applied. The underbody to be primed and treated to two coats of bronze or anti-fouling paint rubbed down smooth. All exterior bright work to be well filled and finished with three coats of best spar composition, each coat to be well rubbed down before the next coat is applied. The interior of the engine space to receive two coats of white lead paint, light slate color. The cockpit deck to be treated to two coats of deck paint of an approved shade. Cove and name to be sized and gilded with gold leaf.

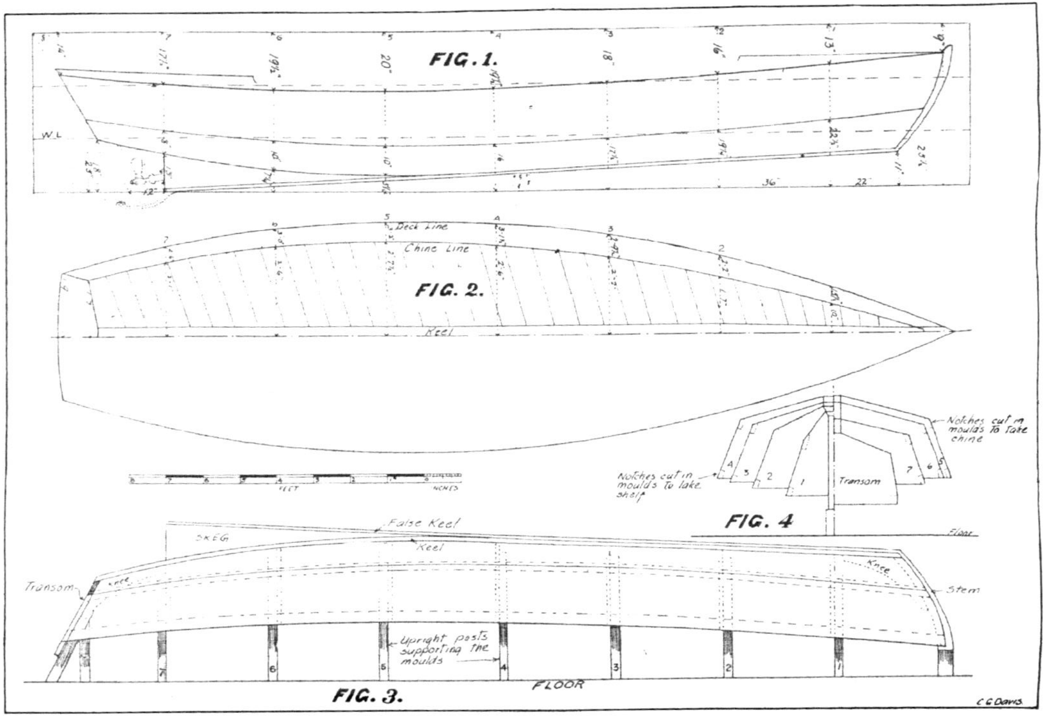

WORKING PLANS FOR A "DEAD-RISE" BOAT

How to Build a "Dead-Rise" Boat

By A. M. KEYS

In response to many recent inquiries regarding the construction of a dead-rise boat similar to the one described in The Motor Boat of July 25th, and in the belief that there are many who would like to have such a craft and who have sufficient mechanical ability to build it, I have endeavored in this article to show a simple and practical method of construction. The boat shown in the drawings is a 30-footer, but can be built with equal ease in any size desired.

First, decide on the size of boat you desire; say, for instance, one of 24 feet over all. Then procure a block of soft pine which will measure 1½ inches for every foot of length of the complete boat, or 36 inches long in this case, and 4½ inches wide by 3 inches in depth. This may be either a solid block or three 1-inch boards tacked one to the other. The latter method is best, since the edges of the boards will offer a straight-edge to cut by and to measure from, the advantage of which will be made clear.

After making sure that your block is perfectly square and smoothed on all four surfaces, if it is a solid block rule it lengthwise into 1-inch divisions. Also divide it into six equal parts by ruling it all the way round every six inches. Then, on one side, draw in the sheer-plan, or side-elevation as shown in Figure 1. On the side which is to be your deck, draw your half-breadth plan. (Figure 2.) Make sure, of course, that the inside of your half-breadth plan is the opposite side from the one on which you have drawn your sheer plan. In order to get fair, sweeping curves make a straight-edge of soft, clear pine, ¼-inch square, and long enough to reach from end to end of your block. Dot off the heights of your gunwale at all the stations and drive pins in at these points. Push your straight-edge up against the pins, and if the curve is fair, rule it in. If not, move any pin up or down until the line makes a fair curve. This is of greater importance than the exact measurements. Decide upon the depth of your side in the same way, making it about 1 inch high for every 12 inches in length, and the ends slightly less, as in the drawings.

Now, having gotten these lines to where they please you, you are ready to cut. First, cut away the block to the line of your keel; then bevel off from that to the bottom of your sides, and you will find that these angles will take care of themselves if your rule will lie fair from bottom to keel at any point. Next, cut your shear, or top of side. Then, having replaced any lines you have cut away, cut your out-board profile, making your angles more acute at the ends than in the middle, and again replace the lines. The block has now only one straight edge, which will be the center line of your boat, fore and aft. Now, lay it down on a piece of heavy paper or on a smooth board of the original size of the block and draw its profile all around, also drawing in the stations. You will need this to get the heights at which your moulds are to be set from the floor, as we are going to build this boat up-side-down, in which position you can get at her best and use your muscles to advantage.

Saw the block into 6-inch sections, where you have drawn in your stations. If the cuts are carefully made, right on the line all round, you will find each cut gives you the shape of half your mould for that station. As we have made 1½ inches to the foot, every ⅛ inch of your block will equal 1 inch of your completed mould. Take off these lines on a piece of heavy building paper, measuring with great care, as a small error in taking off from the block will be just eight times as great in the completed boat. Then cut these out and make your moulds. Any rough stuff will suffice for them, as they will be knocked out and thrown away in due time. After they are complete, notch them out at gunwale and corners, 1½ inches by 4 inches, to take the shelf and chine, as shown by dotted lines in Figure 3. This figure represents the moulds set up, with keel, stem and stern in place, and ready for chine and shelf.

We are now ready to build the boat, and to that end we will follow the advice of the King of Wonderland: "Begin at the beginning, go to the end and then stop." This is how: Strike a line on the floor of your shop with a chalked line, slightly longer than your boat is to be, and right over which the keel is to be laid. Mark this in deeply with pencil and lay out at right-angles the places the moulds will occupy. On these cross lines, nail down pieces of 3x4 inch wood. Rough stuff will do, but put them down strongly. Then to these set up the moulds on legs as shown in Figure 3. Get the heights from your board as shown in Figure 1, and brace them strongly, as they must carry the entire weight of the boat. Be sure they are plumb, up and down, as well as at exact right-angles with the keel. It is well to nail a strip along each side of the bottom, from end to end, to make sure they do not slip during construction. Then set up an upright in front of where your stem will come and also one aft of your stern, right in the centre.

For your keel you will need a piece of oak nearly as long as your over all length, and at least 2x6 inches. Taper this off to 2 by 3 inches for about 6 feet back; a straight cut will do. For the stern, you will require enough 1-inch oak (or 1½ inch if you are going to build a larger boat than 24 feet long) to make it. Don't use ploughed or grooved stuff, but make good, smooth seams, and screw backing pieces on it (not less than three). Don't have a seam run out right at your corners. For the stem a 3-inch oak plank is required. Get the length from your model, and the wider your plank is, the greater curve can be given to the stem. This curve is largely a matter of taste, but as this boat is rather angular anyway, a well curved stem and high crowned deck will add much to her appearance. In getting out your keel, if you have to splice to get the length, let the splice come as far forward as possible, and back it up with a piece of the same stuff, not less than four feet long and bolted through and through. Add stay pieces up and down every three feet to stiffen the sides, and back the joints with oak blocks riveted through; the stay pieces should be jogged over chine and shelf. Taper off the face of your stem and cut the rabbet for the plank ends before setting it up. The keel, for about 18 inches aft of the stem, will also need rabbeting. Let it run off gradually at about that point. The exact angle of the rabbet doesn't matter, as you can doctor it later, when you come to plank.

Now set up the stem and brace it from the upright already mentioned, and from which you also measure the rake of the stem. Line it up carefully with the after upright. Do this carefully, as it will ruin the looks of the job if it isn't plumb. Then saw off the bottoms of your moulds to the width of your keel (6 inches) and brace the keel from above firmly down on your moulds, and also toe in a screw from underneath on each mould into the keel. Then get out your stem-knee (see Figure 3) and bolt or screw it to both keel and stem. Set up your stern in the same manner, getting the rake from the after upright and sawing off your keel so that the face of the stern is outside of it. You can make a pattern for your stern knee after this is done, of any light stuff, and make the knee of 3-inch oak. Bolt it through the keel and screw through the stern.

You can now get out your sides, of 1-inch pine, or, if you can get it, use California red-wood, which is free from knots and comes in long lengths. If you have made your sides flare about as in Figure 4, you will find that your sides will fit the moulds so neatly as to require little cutting at the top, though they will need some at the bottom as that is put up. However, you will find it easy to get at and you can do it quite as well after it is on as before. For the bottom, use 1½ inch stuff, beveling the first plank off (inside) to an inch thickness where it takes the rabbet. Your first few planks will run nearly up and down, but by making them a little narrower at the bottom than at the sides, you can soon get them raking aft. Bevel off keel and chine till the plank lies fair and and fasten with galvanized nails or screws. After the bottom is all planked, plane off the ridge which the plank ends have formed forward, and in fact, all along, if you have beveled your keel so much that the plank ends have met. You will find, by the time you get this far, that you will not need to cut so much bevel unless you want to. Your skeg and your shaft-log will set better if you have left your keel bare for 3 inches in the centre. Make a pattern for the skeg of light stuff and fit it carefully, getting the rough measurement from your sheer plan. Your shaft-log will have to be a piece of 3-inch oak, and after you have gotten it out and drawn a line on it at the angle of your shaft, you will do well to take it to a saw-mill, if not too inconvenient. To bore a hole three feet long, straight through a 3-inch plank is no easy job. If, however, you must do it yourself, get what is known as a barefoot auger, as none other will run as straight for that distance, and fasten with lag screws through keel.

After you have your skeg in place, run your false keel, 1½ x 2 inches, laid flat, clear to the stem, which should have been left long enough for it to butt against. After this, a half-round strap or bang-iron should cover the stem and extend four or five feet back on the keel. Your rudder and propeller guard can best be procured from a dealer in marine hardware and should be of galvanized iron, as, in fact, should be every nail and screw in the boat.

We are now ready to saw off the legs of the mould and to turn her over with the moulds still in place. Build a strong locker across the middle, at about the height you intend your seat to come. This will not only provide a roomy and useful locker, but also will greatly stiffen your boat. The forward ends of both your chine pieces and shelf should be fastened together with triangular blocks of oak or breast-hooks of natural growth and your stem secured to both. Fasten them aft to the stern with oak cleats in the same manner. This, however, is best done before you commence to plank. Fit your deck beams (1x3 inches will be strong enough) one inch below your sides and spring in a 3-inch plank-sheer all along fore and aft.

The moulds can now be taken out and a wide king-plank of oak fitted down the centre of your forward deck, and the rest planked in with pine. Thus far, nothing in the boat has required steaming, and if you have no steam-box you cannot bend your combing. But you can fit one quite as good, if not so shapely, by cutting your forward deck in a V shape and fitting your combing board to it, jogging it out to go over your plank-sheer and springing your combing cold on the sides of plank-sheer to butt up against the forward V. This calls for a nice fitting in the centre, but it can be done, and if finished with a breasthook inside it will look right and be right.

A word about caulking. These boats are generally made without any, and if your joints are carefully made you will not need any. Don't be alarmed, even if you can see the daylight through her before she is launched. Fill your seams with a putty of white lead, to which you have added whiting, worked up to the proper stiffness with a putty-knife. The boat will swell up tight, all right. If you do need caulking, however, hire a professional to do it, as proper caulking is a trade by itself, and many a good boat has been ruined by too much caulking.

Fore and aft bearers 8 to 10 feet long should be laid for the engine. Oak pieces about 2 x 3 inches will be sufficiently strong. Over these your bed-pieces should be jogged. You will have to be governed by the height of your fly-wheel for these, but if you have managed to get this far safely, you can be trusted to proceed unaided.

If your work has been done carefully, you will find that you have a strong, able boat that can go out in any reasonable weather and which can hold her own with any boat of her size and power.

How to Build A Stern Wheeler

BY C. G. DAVIS

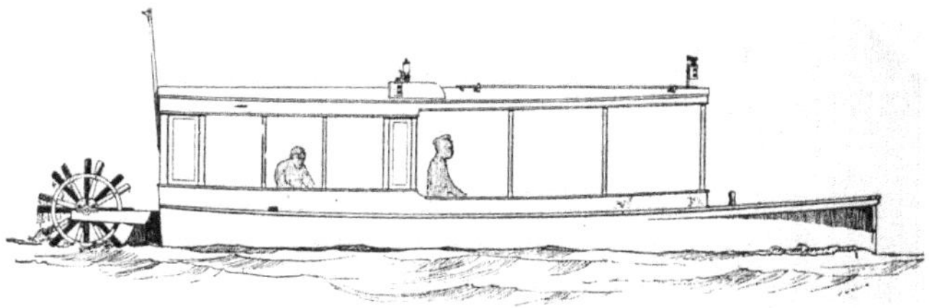

THERE are hundreds of places on the shallow lakes and rivers throughout this country where the sea-going motorboat would be hard aground most of the time, and where only such boats can be used as are designed to navigate the shallow places.

When Eastern designers turn out a Florida cruiser they consider that they have designed a shallow draft boat; but while they consider 36 inches shallow, the places we have in mind would float only 12 inches and in some spots not even that much, places where a man can remove his footgear, roll up his trousers and walk ashore at almost any point.

The enjoyable sensations that make a boat ride so attractive to the people along the seacoast, where motorboating is carried on on a large scale, is all the more of a novelty and all the more enjoyed by those who have never or seldom been on the water on anything other than a log raft. Think what a novelty such a craft would be to a community whose rules of the road could be summed up in two words, "nigh" and "off," or "gee" and "haw." I can just imagine such a crowd with a man at the wheel chewing the end of a dry straw, shouting as he meets another boat, "Gee there! gee!" Why should such a man adopt an unknown vocabulary and say starboard or port? The Government inspector would never come around in such localities to see if the boat had her full equipment of side lights, bow and stern lights, fog horn, bell and life preservers. Who would want a life preserver when the water is only knee deep to a child? And the moo of an old cow in a lot would be as good a lighthouse fog horn as would be needed if the cow was only tied so she'd stay in one place.

There are many disadvantages that the deep water sailor has to contend with that would be unknown on the waters where this stern wheeler could navigate, and no doubt other novel conditions would crop up—a cow might be cooling herself in the brook and in that case a blast of the whistle might be of use to scare her out of the way. Now that small gasolene motors have been perfected and have come into such use on the farm, the man who runs that motor and understands the operation of it could also run the one in the boat, and for that matter the same motor could turn the churn, saw the wood and be put in the boat to drive it as needed. We deep-sea dogs always speak of a boat as she or her, but we're not going to impregnate the fresh, sweet, drinkable water with salty terms, so shall call her "it."

The boat we have selected as being most suitable for such work is one where the hull of the boat is 25 feet long by 7 feet wide, and only sinks 7 inches deep in the water. Including the sternwheel the total length amounts to 29 feet 6 inches.

Any amateur who has never seen a boat would be able to construct a boat this size. Every difficult feature in boatbuilding has been purposely done away with, and any one who can build a box can build it. Two 16-inch planks, 25 feet 6 inches long, 1 inch thick, will make the bottom part of the two sides, and two more the same length, 18 inches deep, will make the top part.

First of all get out an oak stem-piece, 2 feet ½-inch long, beveled off like a wedge, as shown in the detail plan of it on the drawing; then make the transom, as the after end of the box is called, out of 2-inch oak. The shape of this transom is also drawn out, 5 feet 8 inches long by 2 feet 1 inch wide. Nail the lower plank of the sides to the stem so that they make a large Y-shape.

The greatest difficulty in building such skiffs is to get both sides alike and not to have her lopsided, one side flattened and the other bulged out more.

To guard against this, build her on the wooden floor of a barn or shed, so that you can stretch a chalk line and so get a true center line to work by. Square across from this line, where the transom is to go, and measure out the

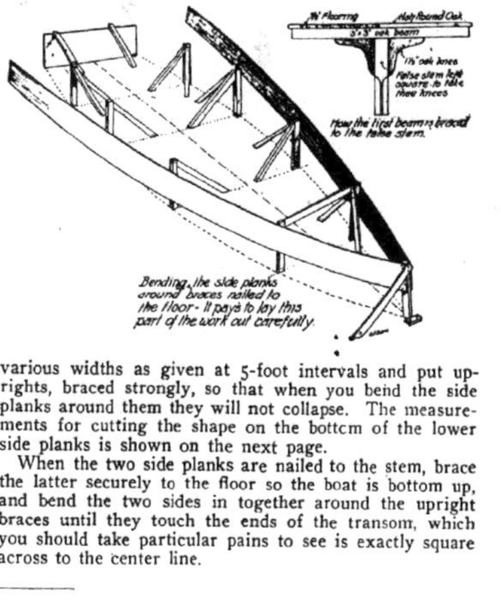

various widths as given at 5-foot intervals and put uprights, braced strongly, so that when you bend the side planks around them they will not collapse. The measurements for cutting the shape on the bottcm of the lower side planks is shown on the next page.

When the two side planks are nailed to the stem, brace the latter securely to the floor so the boat is bottom up, and bend the two sides in together around the upright braces until they touch the ends of the transom, which you should take particular pains to see is exactly square across to the center line.

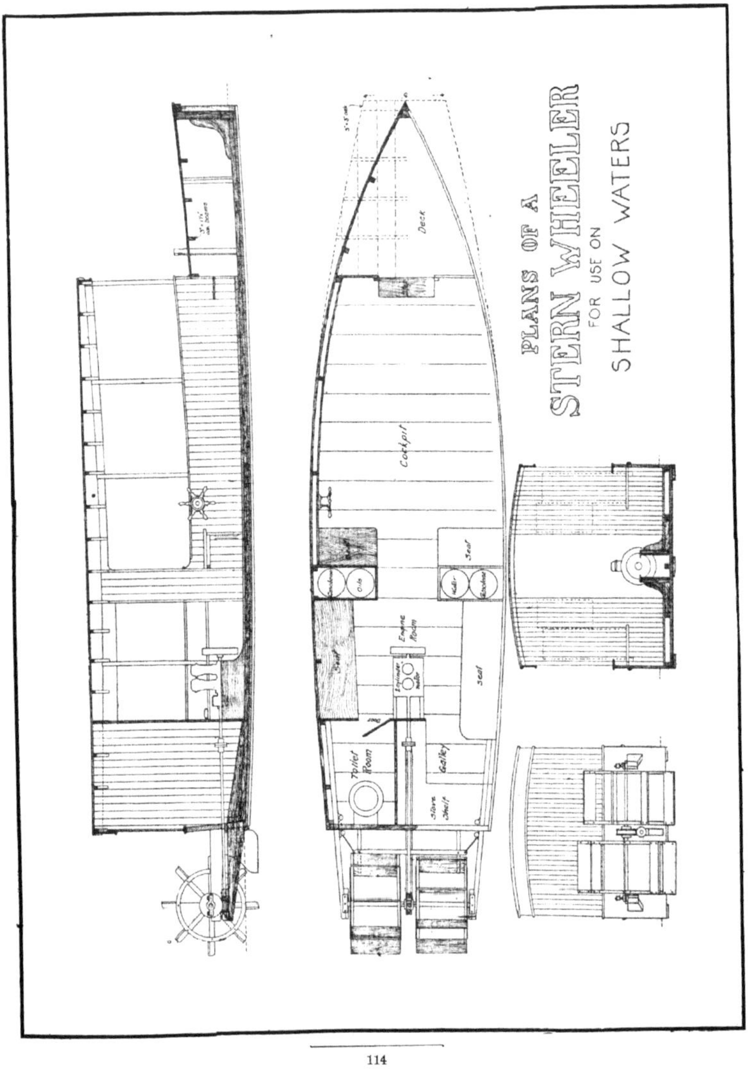
PLANS OF A
STERN WHEELER
FOR USE ON
SHALLOW WATERS
Deck
Cockpit
Seat
Seat
Seat
Seat
Oils
Engine Room
Toilet Room
Galley
Stove Shelf

The sides will not fit flat; they will only touch on one edge of the transom ends, so cut them to the bevel that will make a perfect fit before you nail them fast and always remember in boat work to make the seams so that the outer edges are open a trifle more than the inner edges. If it were a box you would try to make the outer edges invisible, but in boat work this is reversed for the reason

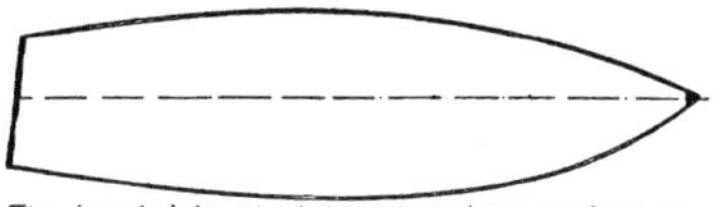

This lopsided boat is what most amateurs produce in their haste to get the boat done. You might better take a little more time and trouble and see that all is square and true.

that putty has to be put in, so that as the wood swells it will squeeze up and be watertight. Some seams will require cotton being tucked in before the putty is applied, where the seams are open a little, and so long as the seams are slightly wedge-shaped, even if the two edges are apart on the inside, the cotton will be jammed in and get tighter and tighter the more it is pushed from tl outside, while, if the seam be the other way, the pressure pushes the cotton through and a leaky boat is the result.

When these side planks are fastened at each end and while the sides are held in true shape, nail a strip of oak, 1½-inch thick and 2 inches deep, so it comes just flush with the edge of the side planks to give a better nailing surface for the nails from the bottom planks. Here is where the friend who has built a box can help. The bottom boards should be about a foot in width and laid across from side to side—lay them close and so the edges make as perfect a seam as possible. Keep the nailheads in, so when you saw them off flush with the side planks you will not ruin the saw teeth by coming in contact with a nailhead. To make the bottom watertight the seams should be calked with cotton, then painted and puttied. Boat cotton is practically the same as cotton batting, only it is strung out into long strings and is of slightly cheaper quality.

If you cannot buy calking cotton take ordinary cotton and string it out and roll it into a cord between the palm of your hand and your knee. After this has been driven into the seams mix up some thin white lead paint and paint each seam to hold the cotton in. When this paint is dry putty the seams, paint the bottom with two good coats of red lead paint and nail on the flat keel and chine, or edge pieces to take the wear and tear off the bottom planks. when the boat lays aground or slides up on a gravel bed. The chine pieces are shown 6 inches wide, but they can be any width, and as three inches will bend around easier some will no doubt use them that width.

When the construction has progressed this far knock the braces out from under the boat and turn it right side up, setting it on logs to raise it off the ground and so as to preserve the proper sweep the bottom is intended to have. Then cut the upper plank to its proper shape and when two of these are ready, one for each side, rivet or nail in the upright posts of 2-inch by 2-inch oak at the various places shown by the measurements in the drawing. Be sure to see that these are all standing plumb before you nail them in.

As most of these boats will be built for use on fresh water rivers and lakes, it is not necessary to use galvanized iron nails or copper rivets as is necessary on salt water, but plain iron nails will do as long as they are kept painted. Punch each nailhead in about an eighth of an inch below the surface of the plank and then fill the hole with putty.

The jogs cut in the top edge of the top plank on the side make the various heights of after cabin, cockpit and forward deck.

To stiffen the bottom planking nail three 2-inch yellow pine stringers forward and aft, which means from front to back in nautical lingo—one right down the middle of the boat and one on each side, leaving a space of 18 inches between them. These should be nailed from the underside of the bottom planks, as they being of pine or cedar, are thinner and softer wood. and the nails, by going

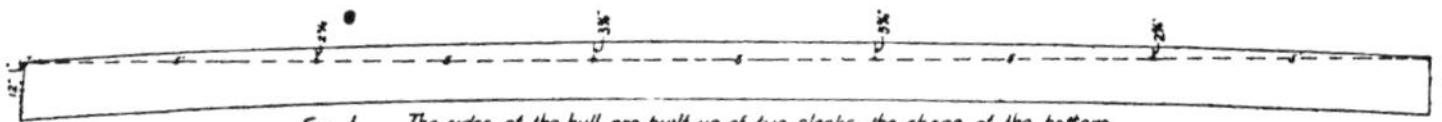

FIG-1 *The sides of the hull are built up of two planks. The shape of the bottom one is shown above. The curve makes it 25 ft 5 ins long. It is 12 ins wide.*

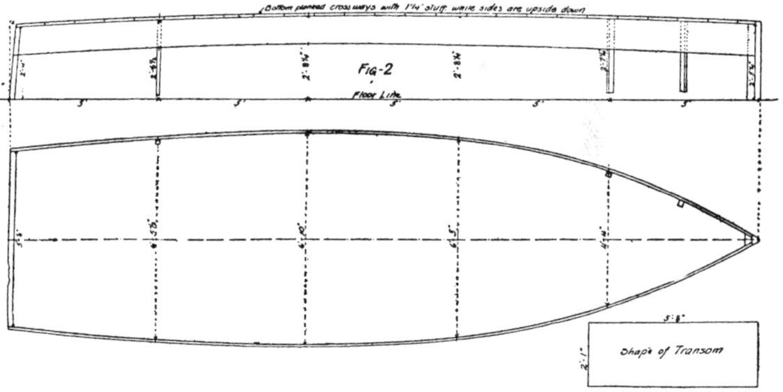

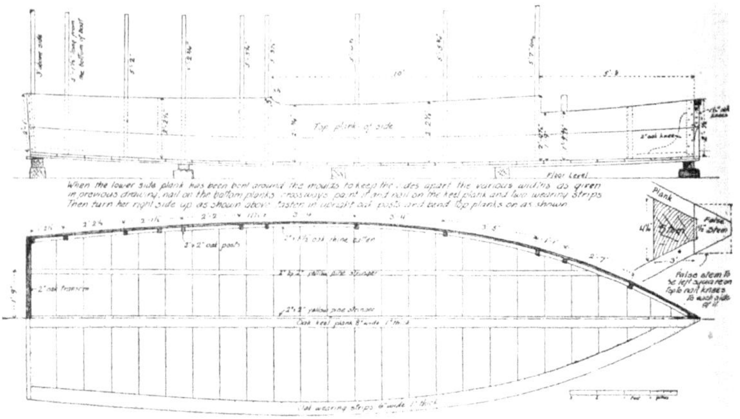

into the harder wood last, will hold stronger than if nailed from above into the soft planking.

If the boat leaks at all, and to keep your feet dry from any rain water that may be in the boat, a flooring can be laid crossways resting on these three stringers and the chine battens at the edges of the sides. To strengthen the corner forward where the stem joins the bottom, as that gets all the ramming up against the shore in making landings, cut a 2-inch thick knee and nail it securely to both batten and inside edge of the stem.

A false stem is then fitted to the real stem. The ends of the side planks and face of real stem can be trued up with a sharp plane, so that the false stem makes a perfect joint —to insure water tightness calk the seams where the sides and stem meet and paint them. The false stem can be pointed off sharp for about one foot up from the bottom. It should be left square from there up, so two small oak knees can be bolted to it to steady the 3-inch by 3-inch deck beam that sets on top of the false stem and against the real stem's face.

Three other beams should then be notched into the top of the side planks, as shown in the plans, to carry the platform deck forward, which is really the front porch, as it were, the place where everybody enters the boat. The platform itself, of about ⅞-inch stuff, is then laid fore and aft over these beams.

The second frame from the bow is to be left sticking up through the deck to form a "hitching post."

To build the roof top, first get out the bands that run around the top ends of the upright posts. These should be of quarter sawed oak about ⅝ to ¾-inch thick, 4 inches deep over the forward cockpit and increased to 7 inches in depth around the after cabin part. The bands across the ends are the same depth and will have to be sawed to shape out of wide planks to get the arch shape, which not only sheds the water as a roof should, but in appearance is far better than flat beams. The beams which go across from side to side are 3 inches in depth by 1¼ inches thick, of oak, and have an arch up of 4 inches. It takes considerably more wood to cut out these curved beams and many men will be tempted to use flat ones, but if this is done the side supports should be carried up higher, so as to give more headroom under the roof.

Nail the ends of the beams to the oak bands and be careful first to bore a hole for the nail or you may split the beam and make a shaky roof.

Over these beams lay a thin roof of ⅜-inch or ½-inch tongued and grooved pine, with the smooth side down paint the top and then stretch wide thin cloth, such as sheeting, over it tacked fast around the edges and laps and then give that two or three thin coats of paint. Cover the edges of the cloth where it is turned over and tacked into the edge of the pine top, with a half round oak moulding about ¾ of an inch wide.

It will add to the appearance of the roof if you also screw or brad on a ½-inch half round oak moulding even with the lower edge of the bands. A similar effect is produced on the top edge of the side planks of the boat by nailing on a cap about ¾-inch thick if it is neatly rounded over on both the inside and outside edges. This cap will be wider because it has to span from the side planks over to and cover up the upper edges of the vertical staving in the cockpit.

This staving and all the bulkheads can be of the same kind of wood, about ½-inch pine, in narrow widths, with its edges tongued and grooved.

An enclosed toilet room is shown built in at the after end. The space opposite may be used as a sort of galley, where simple meals can be cooked on an oil stove.

An arrangement of seats is shown which an owner may change to suit his own ideas. My idea is to leave the forward cockpit practically clear of everything so chairs can be used for seats as they are more comfortable and allow of a change of position when desired.

Two partial bulkheads separate the engine-room, which is aft from the forward part of the boat, and, besides providing a space to carry the gasolene tanks, cylindrical steel or copper boiler shaped ones, and also water coolers, it serves to somewhat deaden the engine noises.

Part II

THE motor installation is simple enough with room to get all around and at the motor, which is set on top of two 3-inch oak beds that are nailed to the floors and braced at each end, particularly at the flywheel end, by two oak knees, which distribute the strain and materially steady the motor against rocking, and yet they need not come so far above the floor as to be stumbling blocks for people passing by it.

The axle of the sternwheel is high enough to enable you to set the motor on a level base, but before you set the motor make your paddle wheel and all its gear, and from this you can get the proper height to set your motor.

To carry the paddle wheels bolt a yellow pine beam 2 inches by 6 inches, 7 feet long, against the inner face of the frames, through a hole chiseled through the transom, so they extend out back of the transom 3 feet. Their outer ends can be lightened up a little by tapering them to a depth of 4 inches at the after end. Rivet on a block slightly wedge-shaped so that its inner face stands fore and aft. This gives a broader landing for the bearings of the paddle wheel shaft. Another stout bearer of yellow pine 8 inches deep by 3 inches thick is mortised through the transom in the center, so it lands right on top of the floor of the boat and to which it is nailed. It is tapered down in height as it goes forward until it is the same depth as the 2 by 2 inch keelson in the center at the after end of the engine. It also tapers up on the after end until it is only 3 inches by 3 inches. This bearer carries the shaft bearings where the worm wheel works into a gear wheel keyed to the paddle wheel shaft or axle.

This worm and gear wheel you will have to buy, and it should be a 4 to 1 gear; that is, for four revolutions of the shaft from the motor the axle of the paddle wheel will turn over once. This, at 300 revolutions and 25 per cent. slip, will give a speed of about 8 miles an hour.

The outer end of the motor shaft will turn in two bearings bolted to the outer end of the central beam, one just in front of and one just back of the worm gear, with ball bearing washers between to reduce the friction caused by the thrust that will come upon these bearings when going ahead or backing. The bearings for the axle of the wheel should be fitted with grease cups on top, and be careful to see that the axle bearings are perfectly level so the wheel will not be lop-sided.

Various other means are sometimes employed for driving the paddle wheel. Some use a sprocket-wheel set-off on one end of the axle, with a sprocket chain connection to a smaller wheel on the end of the motor shaft with the motor set so it extends crossways in the boat. This is a very simple method of transmitting the power, but its objections are the difficulty experienced in taking up the

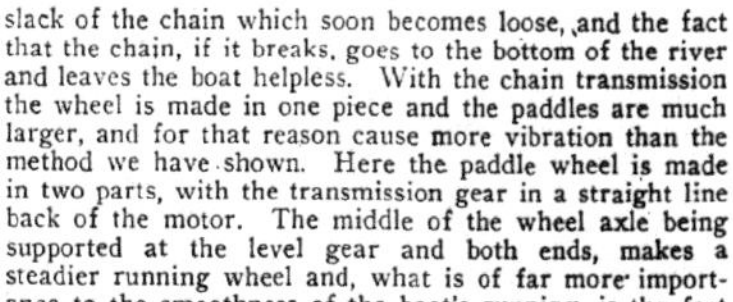

Arrangement of bearings for axle of the stern wheel and end of shaft.

slack of the chain which soon becomes loose, and the fact that the chain, if it breaks, goes to the bottom of the river and leaves the boat helpless. With the chain transmission the wheel is made in one piece and the paddles are much larger, and for that reason cause more vibration than the method we have shown. Here the paddle wheel is made in two parts, with the transmission gear in a straight line back of the motor. The middle of the wheel axle being supported at the level gear and both ends, makes a steadier running wheel and, what is of far more importance to the smoothness of the boat's running, is the fact that the wheel by being in two sections with the paddles set one slightly in advance of the other the shock of the paddle hitting the water is only half as severe as it would be with one wide paddle board clear across, and its action always having one or two paddles submerged all the time.

This is an important feature of sternwheel propulsion,

for if one bucket is just coming out and another just entering, the load on the engine is at that moment very light, and then heavy the next minute as the bucket goes under, producing a jerky effect on the motor and gears.

Another point to be observed in sternwheel propulsion is to have the paddle wheel as large as possible, so that as the paddles come down into the water the boards, or buckets as some call them, will not come down so as to slap the water with a shock, but will dip in as near edgewise as possible.

The wheel in this particular boat is 3-feet 6-inches in diameter, there being eight arms bolted at the center to a circular iron plate 10 or 12 inches in diameter about ¼-inch thick, with a heavy hub that fits on the 2-inch axle, and to which it is fastened by means of an iron key driven into a slot cut half out of the axle and half out of the hub. The arms may be braced out near the buckets by means of an iron hoop bolted to each arm or by oak braces jammed between each arm.

Both wheels may be exactly alike, but when you cut the keyways to hold them on the axle cut one so it sets the

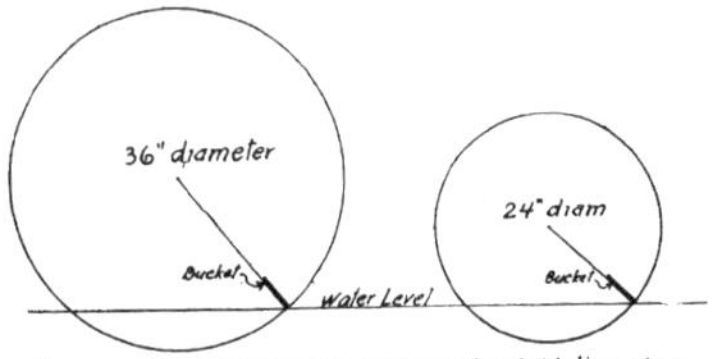

A comparison between a 36" and a 24" diameter paddle wheel - both to dip 4" shows to how much better advantage the bucket on the larger wheel hits the water - less slap and more push.

paddles or buckets about 6 inches in advance of those on the other wheel.

As paddle wheels throw considerable water, they are often partly covered by a shield to keep it from flying about. This, in our case, is taken care of by building up the after end of the boat with a staved up partition that can be made removable all in one piece or in sections to enable you to get at the paddle wheel and rudders whenever any adjustment may be necessary there, and yet it will be tight enough to keep out any water that may splash up against it.

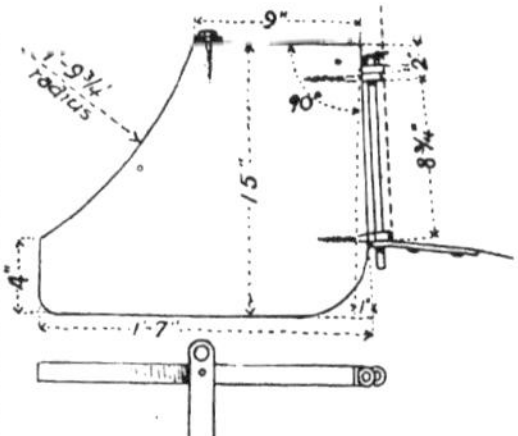

Plan of rudders for Stern Wheeler.

Two styles of rudder hangers - a strap and a screw-eye.

The paddle wheel is kept back from the hull for the reason that the wheel works much more efficiently if it is just on the crest of the stern wave, which will be about where we have located it at the speed she will usually make.

The buckets or paddles should be of hardwood, 21 inches long by 6 inches wide and from ⅝ to ¾ of an inch thick, fastened to the paddle wheel arms by two 5-16-inch carriage bolts into each arm. They should be made as interchangeable as possible, and one or two extra paddles always carried on board to replace any that may be smashed by coming in contact with a piece of driftwood. If all the holes are bored the same distance apart a new paddle already bored for the bolts can be substituted in a few moments.

The steering gear consists of two oak rudders 13 inches long by 12 inches deep, with their after ends cut so as to clear the paddle wheels, hinged to the oak transom, so they drop 3 inches below the bottom and set in 10 inches from the sides so that they will not project beyond the sides when turned around to an angle of 45 degrees.

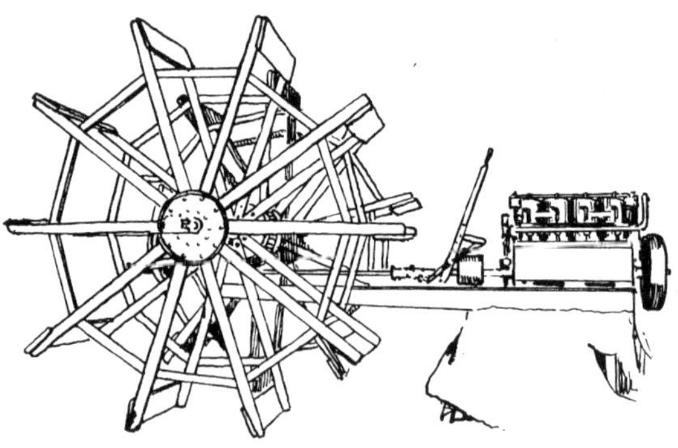

The rudders are hinged by means of two eye bolts bolted through the transom, as shown on the plans, and two eyebolts or iron straps, about 1 inch wide and ⅛-inch thick bolted to the sides of the rudders, and bent around their forward edge so as to form loops, with an iron bolt about ⅜-inch in diameter dropped through them. The upper eye or loop in the rudder, coming just under the eye in the stern, prevents the rudder lifting, and the lower eye, being just above the one in the stern, prevents the rudder dropping. This is as simple and yet as effective a style of rudder hanging as you can get.

By linking the after edge of both rudders together with a flat iron bar 4 feet long, 1½ inches wide and about 3-32 of an inch thick the wheel ropes which lead forward over pulleys or sheaves can be shackled into holes drilled through the outer ends of this bar, and then lead forward to the steering wheel on the lefthand side, so a person can sit on the short seat built just forward of the partition and steer in comfort.

By fitting wire netting screens around the after end of the boat between the roof stanchions, sleeping quarters can be had free from gnats and mosquitoes, and the forward part protected by curtains tacked to the roof band and arranged to roll up when not needed or dropped to keep off the hot sun.

Such a boat will afford a lot of pleasure to a man and his family on waters where an ordinary motorboat could not go at all.

Spotted something that isn't quite right?

Let us know and we'll fix it for future reprints.
And as a thank you for your diligence, we'll
send you a free Sothis book of your choice.

publisher@sothis.press

This page intentionally left blank

Made in United States
Orlando, FL
04 November 2026

90201497R00076